A former New York City police offic... ...iotto had been a fire marshal, an... ...igator, a lieutenant and a captain, ...g made chief in 1992. He holds a B?... ...d has been the recipient of departme... ...nd commendations for bravery and me... ...rvice. Earlier in 2001, he co-authored an a... in *WNYF*, the New York Fire Department's internal magazine, on non-routine Manhattan high-rise fires, that considered some of the same decisions he and his colleagues would have to make at the World Trade Center some months later.

A Staten Island native, Picciotto lives in Chester, New York with his wife Debbie, an obstetrical nurse, and his son, Stephen. His daughter, Lisa, a student at Pace University in Lower Manhattan, viewed the attack on the World Trade Center at first hand from her college campus, and as the towers fell she knew in her heart that her father would be in one of those buildings.

Daniel Paisner is the author of more than twenty books, including bestselling collaborations with prominent politicians, entertainers and businessmen, among them Whoopi Goldberg, Anthony Quinn, George Pataki, Ed Koch, Maureen Reagan, Geraldo Rivera and Montel Williams. He is also the author of *The Ball: Mark McGwire's 70th Home Run Ball and the Marketing of the American Dream*, and the novels *Obit* and *The Full Catastrophe*.

LAST MAN DOWN

The Fireman's Story

FDNY Battalion Commander
Richard 'Pitch' Picciotto

with
Daniel Paisner

ORION

An Orion paperback

First published in Great Britain in 2002
by Orion
This paperback edition published in 2003
by Orion Books Ltd,
Orion House, 5 Upper St Martin's Lane,
London WC2H 9EA

A CIP catalogue record for this book
is available from the British Library.

ISBN 0 75284 941 7

Typeset by Selwood Systems, Midsomer Norton

Printed and bound in Great Britain by
Clays Ltd, St Ives plc.

When I am called to duty, God,
Wherever flames may rage;
Give me strength to save some life,
Whatever be its age.

Help me embrace a little child,
Before it is too late;
Or save an older person,
From the horrors of that fate.

Enable me to be alert,
And hear the weakest shout;
And quickly and efficiently
Put the fire out.

I want to fill my calling,
And give the best in me;
To guard my every neighbour,
And protect his property.

And if, according to your will,
I am to give my life;
Please bless with your protecting hand,
My children and my wife.

The Fireman's Prayer
author unknown

Since 11 September there have been extraordinary efforts to help those who were affected by the tragedy, and I am intent on contributing to this effort.

I will donate a portion of my proceeds from this book to one or more of the following: the Uniformed Fireman's Association Widows and Orphans Fund, the Uniformed Fire Officers Scholarship Fund, the Elasser Fund, and/or the LFF Fund for New York's Bravest, c/o The Leary Firefighters Foundation.

RICHARD PICCIOTTO

CONTENTS

ILLUSTRATION CREDITS

Page ii Big Pictures Ltd.
Page xiii Getty Images. Photographer, Wilhelm Scholz.
Page xv Diagram by Stanley S. Drate.

PICTURE SECTION:
 1 London Features International Ltd. Photographer,
   Dennis Van Tine.
 2 Rex Features Ltd. Photographer, Charles Sykes.
 3 London Features International Ltd. Photographer,
   Dennis Van Tine.
 4 Big Pictures Ltd.
 5 London Features International Ltd. Photographer,
   Dennis Van Tine.
 6 Courtesy of the New York City Fire Department.
 7 Rex Features Ltd. Photographer, Steve Wood.
 8 Rex Features Ltd.
 9 Supplied by Richard Picciotto.
10 Supplied by Richard Picciotto.
11 Rex Features Ltd. Photographer, Jonathan Banks.
12 Rex Features Ltd. Photographer, Jonathan Banks.

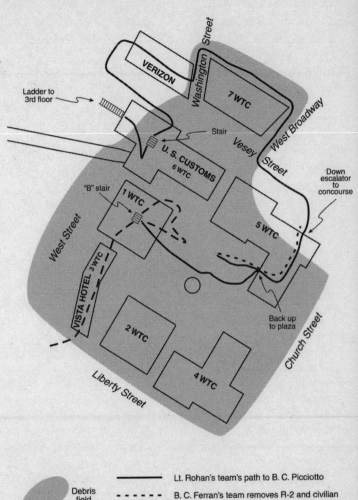

Ladder to 3rd floor

VERIZON

Washington Street

7 WTC

Stair

U. S. CUSTOMS
6 WTC

West Broadway

Vesey Street

"B" stair

1 WTC

5 WTC

Down escalator to concourse

West Street

VISTA HOTEL 3 WTC

2 WTC

Back up to plaza

Church Street

4 WTC

Liberty Street

Debris field

——— Lt. Rohan's team's path to B. C. Picciotto

- - - - B. C. Ferran's team removes R-2 and civilian

– – – B. C. Picciotto's route of the debris field

5-5-5-5

When I first started out, in the early 1970s, it was the custom in the department to sound a sequence of five bells over our internal bell system, four times in a row, whenever a firefighter died on the job. Everyone would stop, wherever they stood, whatever they were doing, for a long moment of silence as the sequence rang out. Five bells, four times.

Each time it came around, it struck me as more bittersweet than the last, because in each ring of the bell was the memory of every firefighter who had died before, mixed in with whoever it was who had died on that day.

Lately, the bell system has fallen out of use, as we've come to rely on different systems of communication, but the signal remains. The call of 'Four Fives', from one firefighter to another, will always signal the loss of a brother. But there is no replacing those bells. I will never forget the sad sound of five bells, four times over, repeated six times after the legendary Waldbaum's fire of 1978, when we lost six good men, and every firehouse in the city went silent as we counted off 120 rings. And I will never forget the bells we never heard on 11 September 2001, when our country was in chaos and our city was in ruins, and 343 of our brother firefighters lay dead in the rubble of the World Trade Center complex. There was no time to ring the bells for these brave soldiers, and too few of us left to hear the ringing.

'Greater love hath no man than to lay down his life for his fellow man,' goes the proverb invoked at countless firefighter funerals and memorials, a tribute that has always echoed in the halls of New York City firehouses through the solemn ringing of those solemn bells.

And so this book is dedicated to the firefighters who gave their lives on that tragic day. They are listed below, 343 strong, by unit and battalion, according to the official list kept by the New York City Fire Department. May their spirits soar, and their legacies linger, and may their mention here stand for the bells that never rang in their honour.

FF	Firefighter	AC	Assistant Chief
LT	Lieutenant	FDC	First Deputy Commander
CPT	Captain	COD	Chief of Department
BC	Battalion Chief	Father	Fire Department Chaplain
DC	Deputy Chief	PAR	Paramedic
FM	Fire Marshall		

Rank	Last name	First name	Unit	Battalion
FF	Anaya Jr	Calixto	ENG004	01
CPT	Farrelly	Joseph	ENG004	01
FF	Riches	James	ENG004	01
FF	Schoales	Thomas	ENG004	01
FF	Tegtmeier	Paul	ENG004	01
FF	Beyer	Paul	ENG006	01
LT	Atlas	Gregg	ENG010	01
FF	Pansini	Paul	ENG010	01
FF	Allen	Richard	LAD015	01
FF	Barry	Arthur	LAD015	01
FF	Kelly	Thomas	LAD015	01
FF	Kopytko	Scott	LAD015	01
FF	Larsen	Scott	LAD015	01
FF	Oelschlager	Douglas	LAD015	01
FF	Olsen	Eric	LAD015	01
FF	Lane	Robert	ENG055	02
FF	Mozzillo	Christopher	ENG055	02
LT	Giammona	Vincent	LAD005	02
FF	Keating	Paul	LAD005	02
FF	Saucedo	Gregory	LAD005	02
LT	Halloran	Vincent	LAD008	02
CPT	Fischer	John	LAD020	02
FF	Cammarata	Michael	LAD011	04
FF	Day	Edward	LAD011	04
FF	Kelly Jr	Richard	LAD011	04
FF	Arce	David	ENG033	06
FF	Bilcher	Brian	ENG033	06
FF	Boyle	Michael	ENG033	06
FF	Evans	Robert	ENG033	06
FF	King Jr	Robert	ENG033	06
FF	Maynard	Keithroy	ENG033	06
LT	Pfeifer	Kevin	ENG033	06
CPT	Brown	Patrick	LAD003	06
FF	Carroll	Michael	LAD003	06
FF	Coyle	James	LAD003	06

Rank	Last name	First name	Unit	Battalion
FF	Dewan	Gerard	LAD003	06
LT	Donnelly	Kevin	LAD003	06
FF	Giordano	Jeffrey	LAD003	06
FF	McAvoy	John	LAD003	06
FF	Ogren	Joseph	LAD003	06
FF	Olson	Steven	LAD003	06
FF	Baptiste	Gerard	LAD009	06
FF	Tierney	John	LAD009	06
FF	Walz	Jeffrey	LAD009	06
BC	Palmer	Orio	BAT007	07
CPT	Farino	Thomas	ENG026	07
FF	Hannon	Dana	ENG026	07
FF	Spear Jr	Robert	ENG026	07
FF	Atwood	Gerald	LAD021	07
FF	Duffy	Gerard	LAD021	07
LT	Fodor	Michael	LAD021	07
FF	Glascoe	Keith	LAD021	07
FF	Henry	Joseph	LAD021	07
FF	Krukowski	William	LAD021	07
FF	Suarez	Benjamin	LAD021	07
FF	Belson	Stephen	LAD024	07
BC	Deangelis (1)	Thomas	BAT008	08
CPT	Burke Jr	William	ENG021	08
FF	McCann	Thomas	ENG065	08
FF	Dipasquale	George	LAD002	08
FF	Germain	Denis	LAD002	08
FF	Harlin	Daniel	LAD002	08
FF	Mulligan	Dennis	LAD002	08
FF	Cain	George	LAD007	08
FF	Foti	Robert	LAD007	08
FF	Mendez	Charles	LAD007	08
FF	Muldowney Jr	Richard	LAD007	08
FF	Princiotta	Vincent	LAD007	08
CPT	Richard	Vernon	LAD007	08
FF	Asaro	Carl	BAT009	09

Rank	Last name	First name	Unit	Battalion
BC	Devlin	Dennis	BAT009	09
FF	Feinberg	Alan	BAT009	09
BC	Geraghty	Edward	BAT009	09
FF	Marshall	John	ENG023	09
FF	McPadden	Robert	ENG023	09
FF	Pappageorge	James	ENG023	09
FF	Tirado Jr	Hector	ENG023	09
FF	Whitford	Mark	ENG023	09
FF	Bracken	Kevin	ENG040	09
FF	Dauria	Michael	ENG040	09
LT	Ginley	John	ENG040	09
FF	Lynch	Michael	ENG040	09
FF	Mercado	Steve	ENG040	09
FF	Gill	Paul	ENG054	09
FF	Guadalupe	Jose	ENG054	09
FF	Ragaglia	Leonard	ENG054	09
FF	Angelini Jr	Joseph	LAD004	09
FF	Brennan	Michael	LAD004	09
FF	Haub	Michael	LAD004	09
FF	Lynch	Michael	LAD004	09
LT	O'Callaghan	Daniel	LAD004	09
FF	Oitice	Samuel	LAD004	09
FF	Tipping II	John	LAD004	09
CPT	Wooley	David	LAD004	09
CPT	Callahan	Frank	LAD035	09
FF	Giberson	James	LAD035	09
FF	Morello	Vincent	LAD035	09
FF	Otten	Michael	LAD035	09
FF	Roberts	Michael	LAD035	09
FF	Casoria	Thomas	ENG022	10
FF	Sabella	Thomas	LAD013	10
LT	Perry	Glenn	LAD025	11
FF	Correa	Ruben	ENG074	11
FF	Barnes	Matthew	LAD025	11
FF	Collins	John	LAD025	11

Rank	Last name	First name	Unit	Battalion
FF	Kumpel	Kenneth	LAD025	11
FF	Minara	Robert	LAD025	11
FF	Rivelli Jr	Joseph	LAD025	11
FF	Ruback	Paul	LAD025	11
BC	Marchbanks Jr	Joseph	BAT012	12
BC	Scheffold	Fred	BAT012	12
LT	Nagel	Robert	ENG058	12
FF	Bielfeld	Peter	LAD042	26
LT	Jovic	Anthony	LAD034	13
LT	Garbarini	Charles	LAD061	15
FF	O'Hagan	Thomas	ENG052	27
FF	Cordice	Robert	ENG152	21
LT	Margiotta	Charles	LAD085	23
FF	Buck	Greg	ENG201	40
LT	Martini	Paul	ENG201	40
FF	Pickford	Christopher	ENG201	40
FF	Schardt	John	ENG201	40
FF	Joseph	Karl	ENG207	31
FF	Powell	Shawn	ENG207	31
FF	Reilly	Kevin	ENG207	31
FF	Derubbio	David	ENG226	31
FF	McAleese	Brian	ENG226	31
FF	Smagala Jr	Stanley	ENG226	31
LT	Mitchell	Paul	LAD110	31
LT	Wallace	Robert	ENG205	32
FF	Henderson	Ronnie	ENG279	32
FF	Ragusa	Michael	ENG279	32
FF	Rodriguez	Anthony	ENG279	32
FF	Byrne	Patrick	LAD101	32
FF	Calabro	Salvatore	LAD101	32
LT	Gullickson	Joseph	LAD101	32
FF	Maffeo	Joseph	LAD101	32
FF	Agnello	Joseph	LAD118	32
FF	Cherry	Vernon	LAD118	32
LT	Regan	Robert	LAD118	32

Rank	Last name	First name	Unit	Battalion
FF	Smith Jr	Leon	LAD118	32
FF	Vega	Peter	LAD118	32
FF	Regenhard	Christian	LAD131	32
FF	Bocchino	Michael	BAT048	48
BC	Grzelak	Joseph	BAT048	48
FF	Coakley	Steven	ENG217	57
LT	Phelan	Kenneth	ENG217	57
FF	Chipura	John	ENG219	57
LT	Ahearn	Brian	ENG230	57
FF	Bonomo	Frank	ENG230	57
FF	Carlo	Michael	ENG230	57
FF	Stark	Jeffrey	ENG230	57
FF	Whelan	Eugene	ENG230	57
FF	White	Edward	ENG230	57
LT	Bates	Steven	ENG235	57
FF	Chiofalo	Nicholas	ENG235	57
FF	Esposito	Francis	ENG235	57
FF	Fehling	Lee	ENG235	57
FF	Veling	Lawrence	ENG235	57
CPT	Brunton	Vincent	LAD105	57
FF	Kelly	Thomas	LAD105	57
FF	Miller Jr	Henry	LAD105	57
FF	Oberg	Dennis	LAD105	57
FF	Palombo	Frank	LAD105	57
LT	Sullivan	Christopher	LAD111	37
BC	Haskell Jr	Thomas	LAD132	38
FF	Jordan	Andrew	LAD132	38
FF	Kiefer	Michael	LAD132	38
FF	Mingione	Thomas	LAD132	38
FF	Vigiano II	John	LAD132	38
FF	Villanueva	Sergio	LAD132	38
CPT	Moody	Thomas	ENG310	58
AC	Barbara	Gerard	CMDCTR	OP
AC	Burns	Donald	CMDCTR	OP
BC	Downey	Raymond	SOC	SOC

Rank	Last name	First name	Unit	Battalion
BC	Fanning	John	HAZMOP	SOC
BC	Stack	Lawrence	SFTYB1	SOC
DC	Kasper	Charles	SOC	SOC
BC	Moran	John	SOC	SOC
FF	Gardner	Thomas	HMC001	SOC
FF	Hohmann	Jonathan	HMC001	SOC
FF	Rall	Edward	RES002	SOC
FF	Blackwell	Christopher	RES003	SOC
FF	Gambino Jr	Thomas	RES003	SOC
FF	Meisenheimer	Raymond	RES003	SOC
FF	Regan	Donald	RES003	SOC
FF	Spor	Joseph	RES003	SOC
LT	Dowdell	Kevin	RES004	SOC
CPT	Hickey	Brian	RES004	SOC
FF	Bergin	John	RES005	SOC
FF	Bini	Carl	RES005	SOC
FF	Fiore	Michael	RES005	SOC
FF	Fletcher	Andre	RES005	SOC
LT	Harrell	Harvey	RES005	SOC
FF	Mascali	Joseph	RES005	SOC
FF	Miller	Douglas	RES005	SOC
CPT	Modafferi	Louis	RES005	SOC
FF	Palazzo	Jeffrey	RES005	SOC
FF	Rossomando	Nicholas	RES005	SOC
FF	Box	Gary	SQD001	SOC
FF	Butler	Thomas	SQD001	SOC
LT	D'Atri	Edward	SQD001	SOC
LT	Esposito	Michael	SQD001	SOC
FF	Fontana	David	SQD001	SOC
LT	Russo	Michael	SQD001	SOC
FF	Siller	Stephen	SQD001	SOC
FF	Cullen III	Thomas	SQD041	SOC
FF	Hamilton	Robert	SQD041	SOC
LT	Healey	Michael	SQD041	SOC
FF	Lyons	Michael	SQD041	SOC

Rank	Last name	First name	Unit	Battalion
FF	Sikorsky	Gregory	SQD041	SOC
FF	Vanhine	Richard	SQD041	SOC
FF	Coleman	Tarel	SQD252	SOC
FF	Kuveikis	Thomas	SQD252	SOC
FF	Langone	Peter	SQD252	SOC
FF	Lyons	Patrick	SQD252	SOC
FF	Brennan	Peter	SQD288	SOC
FF	Gies	Ronnie	SQD288	SOC
FF	Hunter	Joseph	SQD288	SOC
FF	Ielpi	Jonathan	SQD288	SOC
LT	Kerwin	Ronald	SQD288	SOC
FF	Rand	Adam	SQD288	SOC
FF	Welty	Timothy	SQD288	SOC
FF	Scauso	Dennis	HMC001	SOC
FF	Smith	Kevin	HMC001	SOC
FF	Geidel	Gary	RES001	SOC
FF	Marino	Kenneth	RES001	SOC
FF	Sweeney	Brian	RES001	SOC
FF	Weiss	David	RES001	SOC
LT	Martin	Peter	RES002	SOC
FF	Napolitano	John	RES002	SOC
FF	Holohan	Thomas	ENG006	SOC
FF	Johnston	William	ENG006	SOC
FF	Olsen	Jeffrey	ENG010	1
FF	Tallon	Sean	LAD010	1
LT	Leavey	Joseph	LAD015	1
FF	Apostol Jr	Faustino	BAT002	2
BC	McGovern	William	BAT002	2
BC	Prunty	Richard	BAT002	2
LT	Freund	Peter	ENG055	2
FF	Russell	Stephen	ENG055	2
FF	Arena	Louis	LAD005	2
FF	Brunn	Andrew	LAD005	2
FF	Hannafin	Thomas	LAD005	2
FF	Santore	John	LAD005	2

Rank	Last name	First name	Unit	Battalion
LT	Warchola	Michael	LAD005	2
FF	Burnside	John	LAD020	2
FF	Gray	James	LAD020	2
FF	Hanley	Sean	LAD020	2
FF	Laforge	David	LAD020	2
FF	Linnane	Robert	LAD020	2
FF	McMahon	Robert	LAD020	2
BC	Ryan	Matthew	BAT004	4
FF	Heffernan	John	LAD011	4
LT	Quilty	Michael	LAD011	4
FF	Rogan	Matthew	LAD011	4
BC	Williamson	John	BAT006	6
FF	Delvalle	Manuel	ENG005	6
FF	Maloney	Joseph	LAD003	6
FF	McSweeney	Timothy	LAD003	6
LT	Desperito	Andrew	ENG001	7
FF	Weinberg	Michael	ENG001	7
FF	Juarbe Jr	Angel	LAD012	7
FF	Mullan	Michael	LAD012	7
CPT	Brethel	Daniel	LAD024	7
FF	Parro	Robert	ENG008	8
CPT	Clarke	Michael	LAD002	8
FF	Ill Jr	Frederick	LAD002	8
FF	Molinaro	Carl	LAD002	8
FF	Gary	Bruce	ENG040	9
FF	Santora	Christopher	ENG054	9
FF	Elferis	Michael	ENG022	10
FF	Kane	Vincent	ENG022	10
FF	McWilliams	Martin	ENG022	10
FF	Hetzel	Thomas	LAD013	10
CPT	Hynes	Walter	LAD013	10
FF	McHugh	Dennis	LAD013	10
FF	Stajk	Gregory	LAD013	10
FF	Curatolo	Robert	LAD016	10
LT	Murphy	Raymond	LAD016	10

Rank	Last name	First name	Unit	Battalion
FF	Giordano	John	ENG037	11
LT	Guja	Geoffrey	ENG082	26
CPT	O'Keefe	William	ENG154	22
LT	Wilkinson	Glenn	ENG238	28
FF	Cannizzaro	Brian	LAD101	32
FF	Kennedy	Thomas	LAD101	32
FF	McShane	Terence	LAD101	32
FF	Davidson	Scott	LAD118	32
CPT	Egan Jr	Martin	LAD118	32
LT	Suhr	Daniel	ENG216	35
LT	Petti	Philip	LAD148	48
BC	Cross	Dennis	BAT057	57
FF	Leavy	Neil	ENG217	57
FF	York	Raymond	ENG285	51
FF	Cawley	Michael	LAD136	46
FF	Bedigian	Carl	ENG214	37
FF	Florio	John	ENG214	37
FF	Roberts	Michael	ENG214	37
FF	Watson	Kenneth	ENG214	37
CPT	Stackpole	Timothy	LAD103	39
LT	Harrell	Stephen	LAD157	41
COD	Ganci Jr	Peter	COFDPT	
FDC	Feehan	William		
Father	Judge	Mychal		
LT	Crisci	John	HMC001	
CPT	Waters	Patrick	HMC001	
FF	Crawford	Robert	SFTYB1	
BC	Paolillo	John	SOC	
FF	Carey	Dennis	HMC001	
FF	Demeo	Martin	HMC001	
FF	Angelini	Joseph	RES001	
CPT	Hatton	Terence	RES001	
FF	Henry	William	RES001	
LT	Mojica	Dennis	RES001	
FF	Montesi	Michael	RES001	

Rank	Last name	First name	Unit	Battalion
FF	Nevins	Gerard	RES001	
FF	O'Keefe	Patrick	RES001	
FF	Lake	William	RES002	
FF	Libretti	Daniel	RES002	
FF	O'Rourke	Kevin	RES002	
FF	Quappe	Lincoln	RES002	
FF	Foley	Thomas	RES003	
FF	Schrang	Gerard	RES003	
FF	Farrell	Terrence	RES004	
FF	Mahoney	William	RES004	
FF	Nelson	Peter	RES004	
FF	Pearsall	Durrell	RES004	
FF	Tarasiewicz	Allan	RES005	
BC	Amato	James	SQD001	
FF	Carroll	Peter	SQD001	
FF	Garvey	Matthew	SQD001	
FF	Allen	Eric	SQD018	
FF	Fredericks	Andrew	SQD018	
FF	Halderman	David	SQD018	
FF	Haskell	Timothy	SQD018	
LT	McGinn	William	SQD018	
FF	Mojica	Manuel	SQD018	
FF	Virgilio	Lawrence	SQD018	
LT	Higgins	Timothy	SQD252	
FF	Prior	Kevin	SQD252	
FM	Bucca	Ronald	MNBAS	
PAR	Quinn	Ricardo	7990	
PAR	Lillo	Carlos	7981	

11 SEPTEMBER 2001: 9:59AM

It came as if from nowhere.

There were about two dozen of us by the bank of elevators on the 35th floor of the north tower of the World Trade Center. We were firefighters, mostly, and we were in various stages of exhaustion. Some guys were sweating like pigs. Some had their turnout coats off, or tied around their waists. Quite a few were breathing heavily. Others were raring to go. All of us were taking a beat to catch our breaths, and our bearings, figure out what the hell was going on. We'd been at this thing, hard, for almost an hour, some a little bit less, and we were nowhere close to done. Of course, we had no idea what there was left to do, but we hadn't made a dent.

And then the noise started, and the building began to tremble, and we all froze. Dead solid still. Whatever there had been left to do would now have to wait. For what, we had no idea, but it would wait. Or, it wouldn't, but that wasn't the point. The point was that no one was moving. To a man, no one moved, except to lift his eyes to the ceiling, to see where the racket was coming from. As if we could see clear through the ceiling tiles for an easy answer. No one spoke. There wasn't time to turn thought into words, even though there was time to think. For me, anyway, there was time to think, too much time to think, and my thoughts were all over the place. Every possible worst-case scenario, and a few more besides. The building was shaking like in an earthquake, like an amusement park thrill ride gone berserk, but it was the rumble that struck me still with fear. The sheer volume of it. The way it coursed right through me. I couldn't think what the hell would make a noise like that. Like a thousand runaway trains speeding towards me. Like a herd of wild beasts. Like the thunder of a rockslide. Hard to put it into words, but whatever the hell it was it was gaining speed, and gathering force, and getting closer, and I was stuck in the middle, unable to get out of its path.

It's amazing, the kind of thing you think about when there should be no time to think. I thought about my wife and my kids, but only fleetingly and not in any kind of life-flashing-before-my-eyes sort of way. I thought about the job, how close I was to making deputy. I thought about the bagels I'd left on the kitchen counter back at the firehouse. I thought how we firemen were always saying to each other, 'I'll see you at the big one'. Or, 'We'll all meet at the big one'. I never knew how it started, or when I'd picked up on it myself, but it was part of our shorthand. Meaning, no matter how big this fire is, there'll be another one bigger, somewhere down the road. We'll make it through this one, and we'll make it through that one, too. I always said it, at big fires, and I always heard it back, and here I was, thinking I would never say or hear these words again, because there would never be another fire as big as this. This was the big one we had all talked about, all our lives, and if I hadn't known this before – just before these chilling moments – this sick, black noise now confirmed it.

I fumbled for some fix on the situation, thinking maybe if I understood what was happening I could steel myself against it. All of these thoughts were landing in my brain in a kind of flashpoint, one on top of the other and all at once, but there they were. And each thought landed fully formed, as if there might be time to act on each, when in truth there was no time at all.

Somehow, in the middle of all this fear and uncertainty, I got it into my head that one of the elevator cabs had broken loose and was now cascading down the shaft above us. It felt like something was approaching fast, and this was what made sense. Of course, it also made sense that a single, falling elevator cab would in no way generate the same kind of all-over, all-consuming noise and rumble, and I had this thought as well, but nothing else figured. When you're caught in the middle of such improbable circumstances, you grab at what you can find, and this was what I managed. Elevator cabs, freefalling one into the other, like dominoes, filling the shafts that surrounded us with unimaginable terror.

Or, something.

But what? What could make such a loud, horrific, thundering noise? A noise that would surely claim me and my couple dozen brothers from the New York City Fire Department, stranded there on the 35th floor of a torched landmark that had been attacked by a hijacked 767 just an hour earlier. What else could be gaining on us with that kind of ferocious velocity?

It all happened in an instant, and yet it was an instant frozen – iced over to where we all had time to sort through our worst fears, rejecting the ones that weren't grounded in anything real and accepting the ones that seemed plausible. It was a mad race to imagine the unimaginable, to think the unthinkable. Really, I had no idea. I had no idea, and I had every idea. And I stood there, by the windowless elevator banks of the 35th floor, alongside my brother fire-fighters, staring at the ceiling, waiting for it to come crashing through.

Whatever it was.

ONE: Morning

I remember what we all remember about that morning: clear horizon, high sun, visibility stretching to forever. Looking back, it was the beautiful day that killed us, because if it had been grey, or foggy, or overcast there's no way those bastards could have flown those planes. Not on that day, anyway. All up and down the east coast, it was the same: still winds, blue skies and not a cloud in sight. Boston, New York, Washington DC ... all dawning like a picture postcard. What are the friggin' odds of that?

In our house in Chester, New York, about 60 miles north of the George Washington Bridge, 11 September 2001 started early. God's country – or, anyway, a solidly blue-collar community, solidly embraced by firemen and cops and other civil servants who couldn't afford to live in the city they served. I was scheduled for a straight tour, nine to six, which for me comes around just a few times each year. Most days, I'm working six o'clock at night until six o'clock the next night; a couple of times a month, I'm on from nine in the morning until nine the next morning; and every here and there, I'll pull a night tour, a 15-hour shift from six at night until nine the next morning. These straight tours, though, they're pretty rare, especially when you become a chief, as I had done about nine years earlier. I'll tell you, though, they're always a welcome sight on the calendar, for the way they signal a shift that puts you in sync with the rest of the world. Nine to five, or just about. Makes you forget, for at least one day, how out of whack our working lives really are, set against everyone else's.

Chester is about 70 miles, door to door, from my house to the firehouse on West 100th Street. I had the drive timed to a routine. If I had to start at nine, I usually planned to get to work around seven-thirty, which meant leaving the house at about six. That was pretty much the drill. Most guys, they're looking to do the same, itching to start their shifts, to get into it, and at the other end there are guys who've been working twenty-fours, anxious to leave a little early, so it all works out. You pull the same nine hours – or 15, or 24 – you just start a little bit ahead of the books. It's been this way as far back as I can remember, and I've been at this job 28 years. Always, we're looking to punch in early, and to get a jump on heading home. At the front end, there's something about the call and pull of the firehouse that draws us to it for the camaraderie, the bullshitting, the shared purpose, the frat-house environment ... it's different for each of us, I suppose. For me, the pull has always been about the guys and the job. Or, I should say, about the guys and the jobs – emphasis on the plural, meaning all the different fires we've worked over the years. I love talking about this job or that job, big or small, extraordinary or routine. Whatever fire I missed on the previous shift, I have to hear about it. Whatever fire I worked, I have to tell the tale. As chief, and now as battalion commander, I obviously have to hear about each job in order to file the necessary paperwork and stay on top of things, but it runs to something much deeper. I *have* to hear about it. I need to listen to the guys talking about the jobs, down to the smallest detail and over and over, or talking up whatever it was that took place on our tour. It's like lifeblood. I can hear the same story a million times, if it's a good story, and I can tell a good one of my own a million plus. There are a lot of us, caught up in it the same way. It's what we do. It's who we are. And it's why we're there.

So I was looking to leave home early that Tuesday morning, same as my wife Debbie, who worked as an obstetrical nurse at St Luke's Hospital in Newburgh, New York, and my son Stephen, who attended a private Catholic school in New Jersey. My daughter Lisa was a senior at Pace University in Lower Manhattan, so she was the only one in the family likely to be sleeping in at this early hour. We were all out of the door by six, six-fifteen, which meant there was no such thing as breakfast. Maybe a doughnut fisted on the fly. No such thing as small talk, either. Our main thing was getting Stephen moving in time for school. He's always ten, 15 minutes late, for everything, and that's with his mother and me riding him. It's funny the way two people from the same family, the same gene pool, can be so completely different – me, racing to get to work early; Stephen, being pushed along so he isn't too late. And then there's Debbie, in the middle, minding us both.

On this morning, I grabbed a cup of coffee in a travel mug, and did my piece chasing Stephen out of bed and into the shower. We'd never been overly demonstrative with each other, me and Stephen, just kinda nodded or grunted goodbye as we got ready for the day. Two ships passing, that's how it was with our comings and goings each morning. Me and

Debbie, though, we usually kissed, or hugged, and on the days we didn't think to kiss goodbye we usually hollered something to each other on our way out the door. Something nice. Debbie's always told me that when I go off to work there's a part of her thinking about the dangers I might face there, about the job, but she tries to put it out of her mind. Like the wives of most firefighters, she chooses to think I spend my days playing ping-pong at the firehouse, or whipping up great meals with the guys in the kitchen, and we'd fallen into this silent routine where I didn't talk about how close we came in this fire or that fire, how big the job was, or anything like that. She tells me that when I come home at night and crawl into bed alongside her she can still smell the smoke – the fire, even – coming off my body, even after I've showered, but we don't talk about it. We never talk about it. Most guys I know, they don't talk about these things with their wives. The unspoken fear, the unacknowledged fear, is that I might not come home, but it's a fear so ingrained it's almost unnoticeable. It's there, but it's not there, Debbie tells me – back of her mind, but so far back it hardly registers – and knowing this, we try to put our busy routines on pause long enough for a loving goodbye of some kind. This, too, we don't always think about. It has become so ingrained we don't even notice it. We mean to make this moment of connection, but sometimes we forget, or sometimes the clock gets in the way. That's one of the pieces of this morning's ritual that will haunt me later, the not remembering if we had a chance to hug or kiss or say something nice to each other. If it was one of those mornings where we just took each other for granted and went about our business, or if we stopped and paid each other some warm attention. I've reworked the scene a thousand times in my head, and I've got no idea. One minute we were taking turns putting Stephen through his slow paces, and the next minute I was gone.

Gone, first, for bagels. One of the unwritten rules of the fire department is that the guys working the day tour are expected to bring breakfast. It's understood. Chief, captain, lieutenant, fireman … rank doesn't matter. Every day tour, every firehouse in New York City, there are whole units coming in, and it falls to these guys to bring cake, bagels, muffins, rolls, fruit. Whatever they want, but it's got to be something, enough for everybody, so I make a pit stop before I hop on to the New York State Thruway at Rockland Bakery for a bag of bagels. The bakery is a bag-your-own, self-service place, so I grabbed a couple dozen, assorted, with at least one cinnamon raisin for yours truly. I didn't put much thought into it, I just grabbed until the bag was full. Everything else, we've got at the firehouse kitchen: butter, cream cheese, coffee, milk. Every payday, we take up a collection, 20 dollars per man, to keep the kitchen stocked. Coffee's by far the biggest expense, we drink that stuff like it's water, but the 20 dollars also goes to buy condiments and staples and toilet paper. (Actually, we do get department-issued toilet paper, but it's like sandpaper; you can still see the woodchips in each square!) The city doesn't pay for any of that stuff, it all comes from us, and it all starts with breakfast.

I drove my beat-up blue 1991 Honda Accord south to the city in no time at all. I'm usually moving against the traffic, and whenever there was a tie-up there were a bunch of different ways I could go. I'd been driving that route for so long it was like the path I'd worn out between my bedroom and the bathroom – the 160 thousand miles on the odometer were the proof – so it was never a problem making time. I knew all the snags, all the pot-holes, all the trouble-spots. And I made good time this morning, even with the regular commuting traffic, and pulled up at the firehouse around seven-forty. Found a good spot right on the street – the police chief's, from the 24th Precinct, which is housed in the same building (I always made a point of taking their parking spaces, for chop-busting reasons) – made sure my valuables were stowed out of view in the black lawn-and-leaf bags I kept for just this purpose, grabbed the bagels and went inside.

There wasn't much doing, still, from the night before, a couple of incidental runs. But there was some paperwork left over, reports to be filed. The chief I was relieving, Bob Holzmaier, was particularly glad to see me. He'd been on since nine the previous morning, and he took one look at me and started figuring which Long Island Railroad train he could catch out of Penn Station. Bob's not one of those 'minute men' you find in some firehouses, those guys who watch the clock and bug out at the first oppor-tunity, but he was anxious to get home, can't blame him for that, and he had the train schedule burned to that place in his memory where he didn't have to consult it for his ride home.

It had been a quiet night, nothing much to report. Once or twice a year, you'll get a night tour with no calls, but there's usually a run or three. Sometimes, it's a run to nothing doing. There's no such thing as a typical night, but there's always something, and when you check the calls throughout the battalion, there's usually something big. As battalion commander of FDNY Battalion 11, I supervised seven companies, along with battalion chiefs John Hughes, Dennis Collopy and Bob Holzmaier: Engine Co 37 and Ladder Co 40, on 125th Street in Harlem; Engine Co 47, on 113th Street in Morningside Heights; Engine Co 74, on 83rd Street on the Upper West Side; Ladder Co 25 on 77th Street on the Upper West Side; and Engine Co 76 and Ladder Co 22, on 100th Street on the Upper West Side, where I was based. If there were no big jobs here, with our two home companies, there were surely a couple across the battalion.

I dropped the bagels in the kitchen, grabbed my cinnamon and raisin, and hopped up to the office, to go over a couple things with Bob Holzmaier. Things move in much the same way, all over the department, all over the city, as one guy comes in to relieve another. It's like a tag team, or a relay race, the way you pass on responsibility to the next guy, bring him up to speed on what's been going on. Change of shift, there's usually three or four guys in the kitchen, shooting the breeze, talking about the jobs, in no hurry to leave. The morning tour in particular, the guys' kids have already gone off to school, and if their wives work they're usually gone by this time too, so there's no rushing home. What you've got, then, are all these extra hands, guys not wanting to leave for fear they'll miss out on something, and often there's nothing to do but twiddle our thumbs, and sip at some coffee, and pinch at a muffin or a roll, and hash over the details of our last big job like it was the seventh game of the World Series. Or, sometimes, we do actually talk about the seventh game of the World Series, or the Monday night football game from the night before. The firehouse is like a second home to most firefighters, and if our real homes are empty there's no better place to be. There's always something to kick around. And, honestly, there's not a firefighter in my acquaintance who will tell you that he'd rather be anyplace else than chewing the fat with his brothers. Looking back on the last job, looking forward to the next. Shooting the shit.

Business as usual, about to be shattered. I was still in my civvies when I told Bob to take off, but I figured if we got a job I could put on my bunker gear over my clothes. You're not supposed to, goes against regulation, but I'd done it before. Two minutes, I'd be dressed, and good to go. There are some chiefs, they won't relieve the chief on duty until they're sitting at their desks with their ties cinched tight, like they're posing for the newspapers. Regulation says we're to wear our work-duty uniform whenever we're on duty: blue pants, white collared shirt, with the chief's epaulets, gold oak leaves for the chief, silver oak leaves for the battalion commander. It's like the military, silver outranks gold, but if I can't find my silver leaves I'll wear the gold ones – that is, if I can find those. What the hell do I care? How I dress doesn't affect my job performance, how I approach each fire, how I look after my men. If a job comes in, I'm ready; it doesn't matter what I'm wearing.

At about eight o'clock, I wandered down to the kitchen, to talk with the guys, to grab another bagel, to put a head on my coffee. In one sense, the day had officially started with me relieving Bob, but it hadn't really gotten going. We hadn't started checking the equipment, or recharging the batteries for our radios and other gear, which for the morning tour would have to wait until precisely nine o'clock. But the entire house was already in change-of-tour mode. Guys were still trickling in, small-talking on this and that, swapping out upcoming tours. Basically, we were waiting on nine o'clock, hoping our first run could hold out until then. And if it couldn't, then that would be okay, too.

Eight-fifteen, I was back in my office, making ready. A chief's office isn't much: two desks, two chairs, two computers – one for the chief and one for the aide. (My aide, for this tour, would be Gary Sheridan, but he was joined in our rotation by Doug Robinson, Super Dave Shaughnessy and Bobby Pyne.) We kept our family photos and other personal effects in our lockers – in mine, there's a pair of Stephen and Lisa, both of them wearing my chief's hat, circa 1995 or so, which I'm guessing is a fairly standard pose for the children of firefighters – but the office itself is a fairly Spartan, bare-bones scene. Nothing that isn't standard issue, or time-shared by the four chiefs assigned to each house – or, occasionally, by a chief from else-where in the department called in to cover. I've done time in almost every chief's office, and I can tell you that everything's pretty much the same, and in this way we all know where everything is, and how everything works. About the only thing to distinguish them would be the view, but in some quarters you can't even catch a window.

I had some paperwork to cover, some calls to return, and some clothes to step into, so I juggled all three as the rest of my men reported to work and in this way I passed the time until the formal change of shift. At about a quarter to nine, a call came over the 'bitch box', our internal intercom, telling us to turn on the television to Channel Seven. Actually, all it said was, 'Channel Seven', but we knew what the message meant. There were televisions all over the firehouse: in the chief's office, in the kitchen, in the lounge area... If we were tipped to something relevant unfolding on one of the local television stations – a fire, or an interview with one of the brass – the shout was straightforward: 'Channel Seven', or 'Channel Four', or whatever.

It was just after eight-fifty in the morning, according to the clock on the office wall. I flipped on the television and turned to Channel Seven and, as I did so, a couple of the guys wandered into the office to see what was going on. And there it was. Our beautiful blue day turned to shit. Our world turned upside down and inside out and all over the place. Our lives changed forever.

I saw what everyone else saw, couldn't have been more than a minute after it happened. The north tower of the World Trade Center, smoking like crazy. Pandemonium at ground level. And I heard the uncertainty and confusion coming from the announcers. At this early point, the talk on the television was that a plane had crashed into the tower, but there was no indication of the size of the plane, no word on the circumstances of the crash, no accompanying video. Just this incongruous picture of a frightening blue sky, clouded with smoke on a late summer morning.

There were three or four of us in my office, and no one said a word. For the longest time, we just stood there and watched, our mouths hanging open in complete disbelief or miscomprehension. We understood what we were seeing, what the implications were in terms of us firefighters, but it wasn't fully registering, if you know what I mean. Finally, I broke the silence. 'Holy shit!' I stammered, and that about said it. I wasn't much for words to begin with, and here none were needed.

Right away, I knew this was no accident. I knew this in my gut, and I knew this in my heart, and I knew this in my head. Usually, I trust the first two when they're all I've got to go on, but when the head's involved I tend to go that way first. And that way, right away, screamed terrorism. There was always the chance it could have been an intentional act by some nut job loon with no agenda, but I knew enough about flying to realize that this kind of thing didn't happen on its own. I knew that in the history of New York City, there had only been one occasion when an aeroplane had crashed into a skyscraper – the 1945 crash of a B-25 into the 78th and 79th floors of the Empire State Building – and that occurred on a dangerously foggy day without the benefits of the sophisticated flying technology currently in place. Back then, pilots flew by a navigational method known as 'dead reckoning'. They picked a spot on the horizon, and kept it in their sights until they got where they were going. But there was no longer room for this kind of accident. The air space above New York City was a 'no fly zone'. The only strip you're allowed to fly is up and down the waterways, east at an even altitude, west at an odd altitude. The 'rules of the road' are such that there's no colliding, one plane into another, one plane into a fixed object on the ground. When a plane hits a building like that, there's no way it's an accident.

Remember, at this early point, there was no indication of what kind of plane had flown into the tower. Nothing to let us know if maybe there was some kind of bomb on board. Nothing but the sad, surreal picture of a bright, gorgeous sky, blackened by smoke and the thought of the number of people who were surely dead as a result, and the horrible ways they surely died. And to a firefighter, the scene looked even more serious than it did to civilian eyes. I actually heard one of the announcers on the television report that the stock market opening would be delayed by a half-hour because of the crash, and I thought to myself, 'What the hell are these people thinking? What are they talking about? Business as usual? We're looking at the biggest disaster in the history of the city.' Just one glance at the television, and I could see we were into an impossible situation, and if it wasn't impossible it was damn close. There'd be no easy way to get to those office workers trapped above the impact. It could be done, but it wouldn't be easy, and it wouldn't do to be thinking about opening the stock market or going about the rest of our lives in the usual way. Not just yet.

I collected all of these thoughts, and a couple million more, all in the space of a few seconds, and the next thing I knew I was on the phone to the departmental dispatcher. I dialed 261 on our internal telephone system without fully realizing what I was doing, knowing only that those images on the television meant people were hurt, and there was a fire to be put out. It was pure instinct, and as the dispatcher's voice came on at the other end all I was thinking was rescue. How do we get those people out of there? How do we put out a fire on such a high floor? What's our next move? I thought back to an article I'd written a couple of months earlier in *WNYF*, our departmental magazine, about non-routine, high-rise fires, and I figured if there was anyone in the department wanting in on such as this it would have been me.

'This is Chief Picciotto, 11th Battalion,' I barked into the phone. 'I was down there for the World Trade Center bombing in '93. I know the building. If you need my services, call me.'

It wasn't much of a call, but it was a call I had to make. As chief, when there's a fire in your jurisdiction, you just go. You're 'first due', which is department-speak for being first due on the scene. You don't have to wait to be dispatched. There's a 'second due' company as well, and they can usually head out without a direct order, but beyond that you have to follow protocol. If every one of us went out to every fire that caught our interest, you'd have empty firehouses all over the city, so we had to follow command, and in this case command was telling us to stay put. I could live with this, for the time being. I wasn't asking in, just offering myself up, telling them my prior knowledge, which maybe someone in charge might think would come in handy.

Of course, there were others with some of the same pieces of prior knowledge, but I figured I'd get my name in there early. See, back in 1993, 26 February, as a battalion chief in Lower Manhattan, I was the second chief on the scene when a rented Ryder van exploded on the B-2 level of the World Trade Center parking garage. The van had been carrying a nitrourea bomb, 1000 pounds plus, with hydrogen cylinders to add impact, and the detonation took out seven storeys, six of them below ground. Six people were killed, and over one thousand were injured, and the resulting rescue made it the biggest job in FDNY history: roughly 45 per cent of the department's on-duty staff responded to the scene, including 84 engine companies, 60 truck companies, 28 battalion chiefs, nine deputy chiefs, five rescue companies, and 26 other special units. The total response was the equivalent of a 16-alarm fire – an unfathomable amount of manpower and equipment. I was part of the team that established a sub-command post near the entrance to the building, to direct the overwhelming flow of responding units, and I felt strongly that this experience would be useful for whatever the hell was going on in the north tower at just this moment. Plus, I knew the building. I knew the stairwells, and what it was like to evacuate tens of thousands of people through such a narrow space, against an uncertain clock.

In this, I realized, I was like hundreds of other guys still on the job from 1993, but to me it was personal. I hadn't realized it until this black moment, but I felt a connection to those towers. It was the strangest thing. There'd been hundreds of calls to the World Trade Center since it had opened in 1970. Most of these were minor fires, or false alarms, but it was the bombing in 1993 that was the fixed line to where we were this morning. I'd been there in a time of crisis, and I felt I needed to be there during this one as well. I know it sounds crazy to put it in this way, but it's like it was between me and the building, what was happening on these television screens. Like I said, to me it was personal.

I got off the phone and scrambled down to the kitchen, to watch the news with the other men.

Remember, it was still change of tour, so there were guys at both ends of their shifts standing around the television set, wanting to be let in, just like me. Other than Bob Holzmaier, there wasn't a single other guy who had left from the night tour, so the house was pretty crowded. And all of us were just staring, rapt, taking it all in, trying to make sense of what we were seeing, when in truth it made no sense at all. Think about it: how are you supposed to puzzle together images like this into any kind of logical big picture?

Gradually, we started to talk. We could see the amount of fire. We could see the smoke. We could see the hole at the point of impact. We knew this was bad, real bad, that the people above would surely die, if the smoke hadn't gotten to them already. There was no way to get them out. Billy Reynolds of Engine Co 76 mentioned that a lot of firefighters were probably going to die today, and no one could argue the point. In fact, we let his comment just hang there in the room for a while, none of us saying anything, each of us lost in our own private thoughts, wondering which of our brothers we were about to lose, if it would be one of us.

At this point, we were all thinking the smoke would be the killer. We didn't know it had been a huge jet plane, laden with fuel. We all knew how jet fuel could burn at 2,000° Fahrenheit, and how steel can begin to lose its strength at 1,500° Fahrenheit, but we weren't thinking along these lines just yet – and even if we were, we'd remind ourselves that the steel used in the twin towers of the World Trade Center had been coated with some kind of fire retardant. All we knew for certain was what we could see, and what we could see was smoke. Lots of it. Black and billowing and out of control. And we measured what we could see against what we already knew: 90 per cent of all deaths in high-rise fires can be directly attributed to smoke inhalation, so this was a serious race against time. From the pictures we were getting on the television, it didn't even appear that the roof was offering any fresh air, that's how thick the smoke seemed to be, spilling up over the top of the building.

We all watched in horror, and as we started to talk it became clear that we all wanted to get down there, as quickly as possible. We were like racehorses, waiting to be let out of the starting gate. Better, like bucking broncos, busting to get loose. We all felt we belonged down there, for our own reasons, but at the other end of each individual reason was our shared purpose: this was what we did; this was who we were; this was why we were there.

'Chief Pitch,' I'd keep hearing, 'get us down there.' ('Pitch' has been my lifelong nickname, and when we were on duty the guys always put the 'Chief' out in front, mixing the formal with the informal.)

'Fuck command,' I'd also hear, 'we're going.'

'They're gonna send us eventually, so we might as well go now.'

And then the second plane hit. We all saw it coming, couldn't believe what we were seeing. Really, it was such an inappropriate image, it took each of us a beat or two to process it. Who the hell has the tools, or the frame of reference, to understand a sight like that? It was just a beat or two, and yet it felt longer when we were in its middle. We were all silent again, and as each of us in that kitchen began to realize what had just happened, I reached again for the departmental phone. Again, I did this without a clear thought. It was maybe ten seconds from the second crash. I dialled the same number – 261 – for the Manhattan dispatcher in Central Park. I knew they'd be swamped with calls, but I wanted to get mine in there ahead of the others. And it wasn't just departmental calls they'd be dealing with, but 911 calls from people on the scene, cell transmissions from folks trapped on one of those upper floors, nuisance calls from people simply wanting information.

I knew they'd be swamped, but I didn't care. I was just as much of a racehorse as the men in my command. They wanted to be down there, and I wanted to be down there, and if I waited on orders we'd never get anywhere.

'This is Chief Picciotto,' I said again, when the dispatcher picked up. '11th Battalion. A second plane just hit the second tower. Whatever alarm assignment you have, double it. I'm on my way. I know the building, and I'm going.'

Surely, I initially thought, I wasn't the first this guy was hearing of this latest development, but from the moment's hesitation in his voice I started to think perhaps I was.

'Go,' the dispatcher finally said, his voice choked with horror and disbelief and resignation. 'Go.'

I wasn't asking, and I wasn't really waiting for an answer. If the dispatcher had told me to stay put, perhaps I wouldn't have heard him. Sometimes, in the heat of a moment like that, my hearing wasn't so good. But I heard him just fine: 'Go.'

I didn't know it at the time, but my daughter Lisa had bounded from her dormitory on to Fulton Street in Lower Manhattan, to watch these horrific events unfold at first hand, from the relative safety of a few blocks away, and as she stared in disbelief as the second plane staggered into the south tower she realized her father was probably on the scene. She knew it in her bones. That's how a fireman's daughter thinks: you see a burning building, and you know Daddy's there, somewhere, and in this case her bones weren't far off. I wasn't there yet, but I was on my way.

11 SEPTEMBER 2001: 9:59AM, STILL

It was like something out of a Stephen King novel, the way we were all standing there, like stones, waiting for the roar to reach us. A thousand trains. A thousand rushing beasts. A thousand inconceivable terrors, and then a thousand more. Hell, make it a couple thousand, and it still wouldn't have covered it, that's how impossibly loud it was.

We still couldn't know what we were facing, but it was virtually upon us, and we were sure we were about to be pierced or pummelled or pulverized by whatever the hell it was, and I paused in my thinking long enough to frame the scene in my mind: a couple dozen firefighters, spent and anxious and filthy with sweat and smoke and adrenalin, frozen like statues in the corridors of the 35th floor of the north tower of the World Trade Center, eyes skyward, waiting for some unknowable end to burst through the ceiling and overcome us.

Of all the ways to go, I thought, this one would be mine.

I prayed it would be quick. I didn't actually think I was going to die, but I felt certain I was about to suffer. And so I prayed. Quickly.

And yet the roar passed through us, like nothing at all. It was coming, and coming, and then it was on top of us, and a part of us, and finally through us, and now it was gone, rocketing down to the plaza *below. Whoooooooooooshhhhhh! The building kept shaking, our bones kept rattling, but we were still standing. Still standing, and standing still, afraid to move for fear we might upset the uneasy balance of whatever it had been. Whatever might come next.*

No one spoke, for the longest time – and when I use a phrase like 'the longest time', you must realize that everything is relative. This entire episode was only ten seconds in the unfolding, from faint rumbling to deafening roar to fading, falling thunder, but it happened in a weird slow motion. It was eerie as hell, unlike anything I'd experienced, and the strangest, most unsettling part was we couldn't see a thing to correspond to the ridiculously loud noise. It was just noise. Endless, unrelenting, horrifying noise. In a vacuum. We had no idea. Outside, or on television screens the world over, it was clear what we were hearing, but we were marooned in this windowless vestibule, a couple of hundred feet above the ground, another several hundred feet below a raging fire and incredible devastation. And, still, whatever it was hadn't touched us. Whoooooooooooshhhh! It had merely passed through us, and continued on its path, to some unknown place.

Slowly, tentatively, we looked each other over. A few of us were pinching ourselves, to make sure we were okay, and even by pinching we couldn't be sure. Some touched the walls, or leaned a bit harder against them, possibly to make certain the building was still standing, still strong. Some looked around, or over their shoulders, to see their brothers frozen in the same still pose. But no one said a word. Even our radios were silent. We were tuned, variously, to the department's command channel or to our tactical channel, and the odds of both running silent for any stretch were longer than a flagpole's shadow. We always had trouble with our radios, anyway, especially in high-rise buildings, but it was never the case that no one's radio was working for any significant stretch of time. There were maybe 25 or 30 of us, and there wasn't a blip or beep or piece of static among us. It was like we had been dropped on to some other planet, or had slipped on to some other plane of existence. Like our energy supplies had been shut off, and we were powerless to move, or speak, or think clearly.

I don't know how it was on any other floor in the north tower of the World Trade Center at just that black moment, and I don't know that there was another gathering of rescue workers to equal ours in number anywhere in the building. But on the 35th floor, at 9:59 in the morning, I was the one who finally broke the silence. I can't say how it was that I came to speak, or what I was hoping to get back in response. But I was a chief, and I don't think there was anyone else of comparable rank in the vicinity, so it was only fitting that it was me who spoke. I wasn't taking the lead, so much as wondering out loud. And what I was wondering was this: 'What the fuck was that?'

TWO: Ride

A fire chief is only as strong as the aide assigned to him, and on this morning I'd drawn an old pro named Gary Sheridan. Let me tell you something: on a call like this, 130 blocks away, you want a kick-ass driver to get you downtown as quick as possible, and Gary certainly qualified.

Oh, man, did he qualify!

A chief and his aide are always to stay together, in the firehouse and out on a run, and that was how it was with Gary and me. Our schedules matched up, unless one of us had swapped out for a personal obligation, or as a favour to another firefighter of comparable rank, and as soon as I got off the phone with dispatch I sought him out and made ready to leave. The other guys were all wanting in on our assignment, but for now it would just be me going downtown. Me and Gary.

'Come on, Chief,' I'd hear from the firefighters down in the kitchen, 'take me with you.'

'Whaddaya say, Chief Pitch? Got room for a couple more?'

'Hey, what about me?'

But I was going it alone. They cut through me, these pleas wanting in, because I knew I'd be doing the same if the situation was reversed, but this was a matter of protocol. It was also a matter of practicality. I wasn't pulling rank, so much as I was following orders, and there wasn't time to go fishing for more favourable orders. I couldn't simply grab one or two men and head downtown. Things didn't work that way in the department. We had to move as a company, and watch each other's backs as a company, and for the time being the company had not been dispatched. Who the hell knew what the rest of the morning would bring? We couldn't abandon our post, simply because we wanted in on the most hellish fire any of us had ever seen, or because we knew by now our brothers were all over those buildings, and we couldn't abandon our code for the adrenalin rush of the job. The interesting thing about the timing of these attacks was we had a lot of change-of-tour guys, technically off-duty, who could have hopped in with us, or gone down in their own vehicles, and that's eventually what happened all across the department, but I moved about with a kind of tunnel vision. Frankly, I didn't consider this. All I could think was, 'We've got our orders, me and Gary, so let's get the fuck out of here! Let's get moving!' I wanted into those buildings as quickly as possible, and co-ordinating a more comprehensive effort would have just slowed me down.

And so we moved. Me and Gary, in our red and white Chevy Suburban, left out of the firehouse on to 100th Street, right on to Columbus, and a straight shot all the way downtown. Man, I thought, this guy can drive! I knew as much, from our countless calls along endless neighbourhood streets, but I'd never seen him make a run like this! We must have clocked 70, 80 miles an hour, some stretches, and we didn't dip much below 40 or 50. I don't think we even came to a rolling stop, the whole way down. We had our lights and sirens going, and a peeling buzz horn; there was also a microphone in the cab, with a loudspeaker mounted on the rig, in case I needed to shout out at anybody who got in our way, but I kept mostly silent. There were cops at almost every intersection, especially as we reached Midtown and below, and they were doing a great job keeping crosstown traffic in check. A situation like this, you're taking the fire lane, the centre lane, all the way down, but the really good emergency drivers take it like a kind of slalom course. They're swerving left and right, leaning in the direction of crosstown traffic at each intersection, buying themselves an extra five or six feet and an extra beat or two of reaction time in case a car pops out into the intersection. You drive assuming someone's about to bust out into the street at any second, so you're constantly leaning into an anticipated swerve. I imagine it must be a strange sight, to watch one of our high-speed vehicles zigzagging down the centre fire lane for no apparent reason, but there is definitely a reason, and Gary Sheridan was masterful in his execution. I don't think he'd ever been called on to drive like this before, the length of the borough, but he must have done it a million times in his head – and it proved to be good practice.

About the only negative to this kind of driving was the bitch it created for a guy like me struggling to get into his bunker gear on the way. Boots. Turnout pants. Those goddamn suspenders are impossible when you're scrunched into the front seat of a speeding, swerving Suburban, and as I fidgeted into my clothes and banged my head against the passenger side glass I couldn't shake thinking about what we were going to find when we reached the site, what I'd

need to take with me on the way into the building.
Who I'd have to talk to, where I'd be sent, what I'd
be asked to do. My mind was all over the place. I
tried to stay focused, but there was no reigning in my
runaway thoughts, and the radio wasn't helping. We
kept hearing reports of a third plane coming in, and
I was thinking, Strategic Air Command, they had
better be scrambling. I actually saw a low-flying
plane, buzzing down the Hudson to my right, and
assumed the reports were accurate. All air traffic had
been suspended, but there it was, like it was on a
friggin' scenic tour. Either it was one of ours, or one
of theirs. I wondered what the next target would be.
Every possible worst-case scenario was bouncing
through my mind. I thought, 'If I were a terrorist,
I'd go after the George Washington Bridge next.' I
thought, 'You want to send this city into a real
uproar, you take out the tunnels – Lincoln, Queens
Midtown, Holland, Brooklyn Battery – and you really
put this place in utter chaos.' Strategic terrorism.
There'd be no place to go. This was what was racing
through my head, on the drive down. This was war,
and we were off to the front lines. There was a part
of me, deep down, felt like I was driving off to battle,
like I was heading out to do what I had to do, what
needed to be done, but that I wasn't coming back.
Most jobs, this is the furthest thing from my thinking, but here it was, way down deep, in the places I
didn't like to think about. And the weird part about
it, just then, was that I wasn't thinking about Debbie,
or the kids, or my parents in Staten Island, or those
poor people on the upper floors of those towers. I
wasn't even regretting that this was how things were
gonna turn out. I was just fixed on the job, on what

we'd find when we got there, on where we'd move first. I tried to think in details, in specifics. The whole drive didn't take more than eight or nine minutes, but that was enough time to hassle into my clothes and think through every possible situation we might find, and worry what the hell the world would look like at the other end.

When I'd made that second call to dispatch we'd been looking at a four-alarm fire, but I knew by now from what I was hearing on the radio that we were beyond counting. How it works, for the uninitiated, is this: a '1075' signals a working fire, and from there you move to an 'all hands', which means everyone on first alarm assignment is actually on the scene and actively working; a two-alarm fire means you need the full backup, another four engines and three trucks, and it moves up from there, depending on the severity of the situation. On each truck, there'd be a lieutenant or captain and four or five men, depending on the company, so at the four-alarm level we were looking at a whole lot of manpower, and we'd shoot past that soon enough.

As a chief, I considered myself fairly conservative when it came to making these kinds of deployment calls, but my idea of conservative was different to the department's idea of conservative. I was conservative with lives, and not with resources. 'Hope for the best, and prepare for the worst', those were my watch-words, and as we careered downtown I thought back to my blowhard assessment to dispatch, made in heat and haste, to double his alarm assignment. He could double it, and double it, and double it again, and it still wouldn't be enough. What we were facing here was so far beyond the map of our experience in the department, there would be no way to chart it. Hell, I don't know that we could have put out those fires with every damn firefighter in the department. About the best we could hope to do was contain them, save as many people as possible below those fire floors, work the exposures, make sure the whole neighbourhood didn't go up in flames.

When we got down past Chelsea Piers, around 20th Street or so, the roads were completely closed to civilian traffic. We were running along the West Side Highway, nothing but emergency vehicles on the road, and we were gaining on a fire engine that loomed in our path like a road-block. One of the rules of the road in the department is that you don't pass another emergency vehicle unless you're waved ahead by the driver, and here we were, flying towards this engine, about to get stuck in a convoy. You've got to realize, these diesel engines weigh about 400,000 pounds. Tremendous low-end power, but no high-end speed. You get stuck behind one of those babies, you might as well take a nap, but Gary was able to catch the driver's eye in the rear-view, the guy could see it was a chief's car coming up on his rear, so he gave us a little room and we sped past.

We pulled up on West Street, catty-corner to the World Trade Center complex, which was about as close to the towers as we could safely get on wheels, and as I hopped out of the car and caught my bearings I realized this was the exact same spot I parked when I arrived on the scene for the 1993 bombing. There were hundreds of people walking north, in various states of panic and distress but moving in a fairly orderly fashion. It was almost the same scene from the first attack, dressed up in different clothes, and as I surveyed the scene I thought it was like I'd been pulled back, like I'd been cosmically or spiritually bound to take care of some unfinished business, to go through these same motions a second time, and alongside the million or so thoughts bouncing through my racing head I found room for one more: man, this is weird. It goes back to that feeling of connection I wrote about earlier, about feeling drawn back to the building for this second disaster. Like it was some kind of calling. Like it was personal. I looked at the car, parked in roughly the same spot all these years later and thought, 'What are the friggin' odds?'

I reached back into the rig and pulled my flashlight from its recharge holster, and then I quickly found my mask. There were two extra cylinders in the back – compressed air – but a chief's not required to carry a spare so I left them both for Gary. The mask and accompanying single cylinder were heavy enough, and I didn't want to slow myself down for the long trek up those stairs. I wasn't planning to take any tools, wanting to keep light on my feet, but I did grab my bullhorn, and as I did, it occurred to me I'd never used the thing before. I knew from being in those stairwells in 1993, though, that a megaphone would come in handy. If there were ever a couple of buildings built for a rescue worker's bullhorn, it was One and Two World Trade Center.

It was at this point, finally, that I looked up and took in the big picture. The two towers were roaring pretty good. There was heavy, thick smoke on the upper floors, but one glance told me the approaches would be okay. The stairways leading up would be essentially clear, and it was possible the elevators would be working too, at least on the lower floors.

Back in 1993, in the north tower, the smoke conditions were harsh, but entirely manageable if you had experience of that sort of thing. That bomb had been in the cellar, so the smoke was filtering up, and it got to where it was so thick you couldn't see past your own hands in some places. And yet it was just smoke. It was what we call 'oil burner smoke', and it gets you black and filthy and snotty, and you can't scour it out of your clothes, but there's a very high oxygen content to it, which allows you to breathe it in for a lot longer than other kinds of smoke. To a firefighter, a walk through oil burner smoke is like a walk in Central Park. Nothing to it. Now, there are probably 10,000 people out there who'll tell me I'm out of my mind; they'll insist that coming down those stairwells was unbearable, that the smoke was overpowering, but that's a civilian view. I don't mean to diminish what these good people went through, because believe me they went through a lot, but the smoke was manageable that day. That's why so many thousands of people were able to get down safely from the extreme upper floors. If they had been breathing shitty, acrid smoke with a low oxygen content, they wouldn't have made it ten floors. Here, though, below where the planes had hit, there was very little smoke, which was going to make our rescue effort that much easier. At least, setting out, we'd be going up in relatively clean, fresh air.

I took in everything as I stepped from the Suburban, and at the same time I wasn't focused on any single thing, if that makes sense. I saw jumpers, a couple dozen, falling from the sky, desperate to escape the smoke and a suffocating death, and this was the one sight that threw me. I wasn't expecting to see people falling from the sky. There was nothing so desperate about the scene in 1993, nothing so heart-wrenching. I later read that a couple of people, friends or co-workers I suppose, actually jumped hand in hand, but I didn't see any of that. I wasn't focused enough to catch the expressions on these poor people's faces, or what they were wearing, or any personal details, but the mere fact of their jumping was incomprehensible. All that day, all through what was to come, and on into all the days since, I could close my eyes and imagine the faces on these falling bodies, imagine the terror that drove these good people to leap. But at the time, all I could think was, 'Shit, I don't want to get hit by one of these jumpers.' I'm sorry, but that was my survival instinct kicking in. That's how messed up the whole scene was, that I was processing these jumpers as hazards. As debris. I saw falling glass, and furniture, and these struck me in a similar way, at the time. I didn't allow myself the time for it to strike me as sad, the way I grouped all these hazards together, but there would be time for that later, and then it would be sad upon sad upon sad. Man, these poor, desperate people! What the hell else were they supposed to do, on those upper floors? What the hell would I have done, in the same impossible situation? For weeks afterwards, I couldn't think about it, but at the same time I couldn't *not* think about it, if that makes any sense. I don't think I'll ever get my mind around what those last moments must have been like, to push a person to such extremes. I'm not sure I want to. But at just that moment, as I snaked towards the towers, these kinds of thoughts were nowhere in my racing mind. What I was fixed on, totally, was the task at hand – getting inside one of those buildings and racing up to the fire floor – and everything else was a hazard, falling from the sky, keeping me from my goal.

I left Gary Sheridan with the rig and made a quick dash to the north-west corner of the north tower, dodging bodies and glass and falling office equipment along the way. It was like I was running the gauntlet, but I made it in, and I didn't allow myself the thought that sits with me now as I set this to paper: people were *jumping* out of the building, frantic to get out of there, as I was racing to get in. It makes no sense, when you set it out in this way, but I didn't think of it in those terms just then. I didn't think of it at all. Hell, I don't think there was a firefighter on the scene who wanted to be anyplace else but up on one of those fire floors, doing what he could to put out the flames, to save lives.

I should point out here that I wasn't the only fire-fighter concerned about the dangers posed by these jumpers. Almost everyone I talked to who raced into those buildings reported something like the same thought. Civilians too, on the way out, couldn't fathom the sight of all these bodies hurtling towards the ground, and feared once again for their own lives as they scrambled to get out of the way. And, of course, we'd learn later that day that the very first firefighter killed on the scene, Lieutenant Danny Suhr of Engine Co 216, was killed by the body of a woman who leapt from one of the towers. I never learned the woman's name, or what floor she worked on, but a lot of us knew Danny – and all of us knew Mychal Judge, the fire department chaplain who knelt down before Danny to administer last rights. Father Mike took off his helmet to say the prayers, and as he did so he was hit on the head by a falling chunk of the building. His death made headlines and shot through the department something fierce. And so, yeah, it's a hard-hearted way to put it, but it was very clearly a hazard, these people who had no better option but to jump 90, 100, 110 storeys to their certain but perhaps more instantaneous deaths.

. On the ground, on the outskirts of the World Trade Center complex, the scene was somewhat chaotic, but it was a systematic kind of chaos. The police were doing a good job of controlling the situation on West Street, and someone directed me to an opening in the glass wall of the lobby of the north tower. There wasn't time for rescuers to go through the revolving doors by the main entrance on the concourse level, so firefighters had busted out a section of glass, 20 or 30 feet up, as high as they could reach with whatever tools they had, and people were just going in and out of there, not a thought for the still-hanging, un-anchored glass above them. Really, that remaining pane must have weighed a ton, almost literally, and it was probably an inch thick, and there was no longer anything holding it in place but the tension on its sides; it could have slid down at any moment like a guillotine and sliced someone in half, easy, but that was my way in so I took it without a thought. The thought came later, in the rethinking. At the time, I just moved. Fast as I could.

The makeshift command post was just inside the opening, and off to the left. I knew that no matter how crazy things were getting, how frantic everyone was at this point, there's always a method to the madness. There had to be some guy in charge, and that guy was a deputy chief named Pete Hayden. Like

48

I said, this was war, and Pete was our General Schwarzkopf. Pete was one of the great good guys in the department; twice he'd been promoted to a staff position, but each time he didn't like the way the commissioner was running the department or some of the decisions he was asked to make at the expense of his fellow firefighters and he'd asked to go back on the line; he was one of us. There wasn't a trace of brass in Pete's blood. He had a whole batch of companies lined up against the glass exterior wall, waiting to be deployed. There were a few lone wolves rushing into the building without orders, but for the most part the companies arriving on the scene were lined up for assignment. Pete was running the whole north tower command, and he was pretty much on overload. I can't say I would have done any better. He had a couple of dozen people scuttling for his attention, and he was being hit with more information than he could possibly process. There must have been ten other chiefs there, waiting for something to do, and I surveyed the situation and wasn't about to join the line. Command was command, and we were all good soldiers, but I was itching to start moving up to the fire floor. I was caught somewhere between lone wolf and good soldier, but I didn't race down here to stand in line so the wolf won out.

I flashed up and down the line and picked out a group of guys who looked like they were also ready to bust. A truck company, 110, out of Brooklyn. Good soldiers all, racehorses wanting to be let out of the gate. Usually, I scan a group of firefighters and I can pick out a face or two, but I didn't recognize a single one of these guys. I grabbed one of them by the shoulder. 'You guys ready to go?' I asked.

'Yeah, Chief,' he said. 'We're ready.' He didn't have to ask the rest of his company.

'You got all your tools?' I asked.

'Yeah, Chief,' I heard back. 'Got all our tools.'

'Extra cylinders?'

'Extra cylinders.'

I cut the company out of the line and went over to Pete Hayden. I wasn't jumping the line so much as getting something done. Expediting things, that's the way I looked at it. Helping out in whatever way I could. We were now one less company Pete Hayden had to worry about, me and 110 Truck, one less group of men he had to deploy. I hadn't seen Gary Sheridan since I left the vehicle, so it would be these guys watching my back for now – and me watching theirs.

I knew the only way to get Pete's attention was to get in his face, so I walked right up to him. 'Pete,' I shouted, trying to break through the commotion. 'What do you need?'

He looked right through me at first, then did a quick double take. There was a smile of recognition, of shared mission, of mutual disbelief at what was happening. Something. It was like we were both holed up in the same bunker, fighting the same good fight. 'Richie,' he said, 'I got people trapped, office workers, on 21 and 25...'

I didn't need him to finish the thought. I knew the rest. Find these people, help these people, bring them down.

'This tower?' I asked.

'Yeah, this tower. Tower One.'

'You got it,' I said, and turned for the communication co-ordinator, who'd set up his post a couple yards away. There was a command board, opened up just like a suitcase, with a little magnetic sticker corresponding to every company in the department, and it was the co-ordinator's job to keep track of each company's assignment on this board. It was like a portable deployment diagram, and this guy too was on information overload. 'This company goes here. This company goes there.' I'd had this guy's job before, and I didn't envy him, not today. I figured I'd just make his life easier, and mine, so I grabbed my sticker, 11 Battalion, and then I fished for 110 Truck, and I slapped them in place on the board.

'11th Battalion,' I announced to the co-ordinator, as I was reaching past him, 'going to the 21st and 25th floors. I'll be in radio contact with you.' And, for emphasis, I pointed directly to him as I said 'you', so there would be no mistake, so there would be one less thing for this guy to have to remember.

Here again, I wasn't cutting the line, or cutting corners; I was slicing through the bullshit, getting things done, and the entire exchange took about 30, 40 seconds, tops. From the moment I raced through the glass cut-out entry, to the moment I pulled away from the command post with 110 Truck. It all happened before I could think it through. In, assigned and gone – boom, boom, boom – all in no time at all, and yet it shook out pretty much like I'd imagined it, as I'd played the scene out in my head on the drive down.

It was probably about 9:45 in the morning, and I had left my house in Chester roughly three and a half hours earlier – and here I was, finally getting to work.

11 SEPTEMBER 2001: 10:00AM

I got back nothing.

'What the fuck was that?' I tried again.

Still nothing.

All around me, guys were staring at the ceiling, their mouths opened wide enough for their fists. Their eyes big as Frisbees. Their hearts beating like a drum solo. Gradually, we all started to move – slowly, unsurely. But for the longest time, no one said a word. We looked each other over. We looked at the walls and the ceiling and the elevator doors to see that all around us was pretty much as it had been just a couple moments earlier, before the terrifying roar.

I punched the handi-talkie I wore on my chest. The thing had an omni-directional microphone, which allowed you to key it and talk and keep your hands relatively free. 'What was that?' I said into my shoulder. 'Does anyone know what that was? What the hell was that?'

I was operating on the command channel, and figured there had to be someone, somewhere in the vicinity, who could tell me what had made such an impossible noise. But I got back nothing, so I tried it again, more formally this time. 'This is Chief Picciotto, Battalion One One,' I said, for the record. 'I'm on the 35th floor of the north tower. We just had a huge noise in the building. Does anybody know what happened?'

But there was no response.

By this time, some of the other firefighters were keying in on their own radios, searching for their own explanations. They were working the tactical channel, while I continued working the command channel. We were all close enough, and still and silent enough, that we could hear each other's radios, but there was nothing coming back on either channel. No one moved, other than the fidgeting we were all doing with our handi-talkies. No one raced for one of the nearby offices, to look out a window. No one did any kind of spot check of the stairwells, or the surrounding area. We all just stood, mostly silent and still and disbelieving, mostly shifting uncomfortably in place, mostly waiting to be told what had just happened, what might happen next. We were afraid to move, for fear of making the wrong move. And we were desperate for information, for some place to put our fears.

Finally, after the longest couple of seconds, I heard a message on the tactical channel: 'The tower came down.'

That's all it was, and it wasn't in response to one of my questions, or to one of the questions put by one of the firefighters on the floor alongside me. It was just offered, in a tone of shock, for the general consumption of anyone in range: 'The tower came down.'

I overheard the transmission, but it didn't register. You know that line from all those old science fiction movies? 'Error, that does not compute.' That was how I took in this message, at least at first. It made no sense. What tower? What the hell were they talking about? My first thought was that maybe one

*of the radio towers had come down. This seemed
reasonable, given what I'd seen from outside on the
way into the building, what I'd seen on the television.
But a tumbling radio tower couldn't possibly have
made the kind of noise we'd all just heard. Couldn't
have shaken the building and rattled our bones and
passed right through us like a thousand trains. So,
maybe it was a whole bunch of radio towers. Maybe
there had been a couple dozen up there on the roof.
Maybe that was it. But then I rejected this notion and
picked up another. I started thinking it was a water
tower that had come down, and that the rush of sound
had maybe been the tens of thousands of gallons of
water cascading down the building. This, too, seemed
reasonable, but here again the picture I had in my
mind of a falling water tower in no way matched the
sounds I could still hear in my head. The tower came
down. It made no sense. It did not compute.*

*I was still on my own radio, frantically searching
for some kind of confirmation or explanation on
the command channel, but I couldn't raise a signal,
and by now, all around me, these other firefighters
were scratching their heads, trying to figure out what
the message meant. What tower had come down?
Where? What kind of damage did it cause? What's
the fire situation? We were moving now, tentatively,
still essentially in place but turning to each other for
information.*

*To my left, I heard one firefighter remark to
another, or possibly to the group at large, 'The tower
couldn't come down. This is the World Trade Center.
Nothing can bring these buildings down.' Or maybe
he was just thinking out loud, talking himself down
from his worst fears.*

Whatever it was, this was the first time I'd allowed myself to make that kind of connection. This firefighter had put it out there for me, and now I couldn't shake it. The tower came down. The south tower. Tower two of the World Trade Center. The whole building.

The moment I heard it, I knew. We all knew. At the same time, we all knew. There was this weird wave of knowing that washed over each of us, one by one, and before too long it was a part of us, fixed to us, and we stood there, knowing, and not moving, and not knowing what to do.

'Holy shit!' I heard.

'What the fuck?'

'Man!'

'Jesus Christ!'

We were all just talking to hear ourselves talk, making noise because there was nothing to say. My first thought, once the realization hit, was that there had to have been hundreds of firemen in there. People I knew. People I loved. Hundreds of them, easy. Hundreds and hundreds of other people too, maybe thousands, but my first thought was for the firemen. They'd been sprinkled all over that building, same as we were sprinkled all over this one. And while this thought was registering, there came a few more. We'd been hearing all morning that there was a third plane coming, so I started thinking that it was some follow-up attack that had brought the building down. We were also hearing reports of missiles being fired from other high-rise buildings in the area, possibly at these towers. And there was always the prospect of some kind of bomb. We didn't know yet to think in terms of jet fuel or melting steel or anything. We didn't know shit. We were trapped in this interior space, deaf and blind to the rest of the world, to the devastation outside our windows – probably the only group of people in the civilized world who hadn't seen what had happened to the south tower. We might as well have been in a cave, for all we could tell about what was unfolding on the plaza below. We were frozen by a bank of elevators, going on vague dispatches on the radio, standing stock-still while our imaginations ran all over the place.

I closed my eyes and tried to picture what it must have looked like, the entire building coming down, but I couldn't see it. I had no frame of reference for that kind of destruction. All I could see was the aftermath, and not the collapse itself. Open sky, where there had once been a tall building. And all I could think was, 'This is war'. I'd thought it before, all morning long in fact, but here it was again. War. More planes inbound. Missiles being fired from other buildings. Bombs bursting in air. And on and on.

Then I opened my eyes to a new thought: if the south tower could come down, I realized, the north tower could, too. And I knew we had to get out of there.

THREE: Climb

The north tower of the World Trade Center was designed with three separate stairwells – the 'A' and 'C' stairwells at the south-east and south-west corners of the building, and the 'B' stairwell at the centre of the building. The south tower was built the same way, two on the perimeter and one at the core, but my focus here will be on One World Trade Center, also known as tower one, or the north tower.

In the north tower, the only one of the three stairwells to reach the lobby level was the B stairwell. A and C terminated on the mezzanine, one giant floor above the lobby, which meant that the majority of the descending office workers had been exiting there. By all first-hand accounts, those perimeter stairwells were also the most trafficked.

I was on the lobby level with my company of men from 110 Truck, and the cascade of office workers pouring out of the B stairwell was relentless, so I couldn't even fathom what the scene must have been like up on the mezzanine, with the heavier flow from the A and C stairwells. From what I was seeing, though, people were moving in a state of shock, which seemed a much better state than panic, which I would have thought likely. Nobody was screaming. Most were eager and grateful to be directed to safety, and most were patient and keeping their place in line. This, I supposed, was a godsend. Some people were visibly fatigued, and some seemed to be experiencing breathing difficulties, but for the most part people were in good shape. They were helping each other out. They moved briskly but not frantically, swiftly but not hurriedly. There's a difference. No one was running. Folks were quick-stepping their way out of there, but not so fast that they would lose their balance or their composure. I can't say for sure what the scene was like on the mezzanine, with two stairwells of people spilling out on to the floor, and possibly a good percentage of folks from stairway B ducking out at the first opportunity, but I was later told that it too was a methodical procession.

All three staircases were approximately 42 to 48 inches wide. Two people, standing side by side on one of the steps, they'd be touching the walls or railings on the side and touching shoulders in the middle, that's how narrow it was. It was fairly tight, I'll tell you that, and once the firemen and other rescue workers started pressing up, with all of their equipment and bunker gear, you really only had one lane open to downward civilian traffic, so the going was slow and the potential for aggravation and anxiety extremely high. But – and I can't stress this enough – folks were surprisingly composed, and in control of their emotions. We rescue workers were still outnumbered, by a significant margin. There were far more of 'them' coming down than there were of 'us' going up. But that would change.

The stairwells featured what we called 'return' stairs, which meant there was half a flight of stairs zigging one way, a small landing, and then half a flight of stairs zagging back. In other high-rises, you sometimes find 'scissor' stairs, which go back and forth one full storey at a time. This 'return' design was significant, and fortuitous, because it meant that on each floor, at the base of the stairs, we would be positioned in the exact same spot, relative to the elevator banks and the access doors and corridors and so forth. It also offered descending office workers a brief respite on each landing, a small space for someone to catch his or her breath before making the turn and continuing down. Of course, if you had too many people pausing to catch their breath at the same time, on the same landing, it would have bogged things down the same as if they had stopped directly on the stairs, but it was something.

I looked at my guys, and I looked at the slow crawl of hundreds and hundreds of exhausted people plodding down those narrow stairwells, and I thought to myself, 'There's got to be a better way to reach up to the 21st floor'. That was our first destination, and I knew it would take just short of forever going up the clogged stairs. The prospect struck me like walking up a down escalator, against a sea of shoppers headed in the proper direction. I didn't reject the climb because of the effort. I was in strong aerobic shape – I worked out on the stairmaster each day at work, climbing the equivalent of two World Trade Center towers, and I rode my Trek 2300 bicycle four or five days a week, a couple of hours a stretch – but I had no time to spare. Plus, I had about 50 to 60 extra pounds on my back. The guys in 110 Truck, they probably had an extra 100 pounds or so, the main difference being the extra cylinder of compressed air they were made to carry. That, and all of their heavy tools. Me, all I had was my mask and cylinder on my back, and my flashlight and bullhorn in my hands; toss me a second flashlight and I could have juggled the two lights and the bullhorn, no problem.

In a high-rise fire it often makes good sense to use the elevators, if you can determine that they are operating safely, and I knew that in the north tower there were 99 different elevator lines; surely, I thought, one of those shafts had to be clear. Actually, let me restate that. It's not so much that it *often* makes good sense, but it can frequently be a good and viable option. Regulations allow us to take an elevator to five floors below a fire floor and no closer. I had no real idea where the fire floor was in this case, not precisely, but I knew roughly where the first plane hit from those pictures on the television and that it was probably around the 90th floor or so, and I knew the 21st floor would give me plenty of clearance to beat that regulation. By now, we had learned from radio dispatches that these were 767s that had crashed into each tower. Fuel-laden 767s, at the front end of long hauls, so there was the concern over jet fuel. Twenty-two thousand gallons on each aeroplane, it was later reported, and this presented a compelling argument for the haste and efficiency of elevator travel; it also presented a compelling argument against, for the way a resulting fireball might have found us suspended by a thin elevator cable, but these were the types of shift-on-the-fly dilemmas we were made to consider.

I spotted two blue-shirted building maintenance workers stepping out of a working elevator car, and I shot over to them. 'Hey, buddy,' I hollered at the one closest to me, 'how far up does this thing go?'

'Sixteen,' he said.

'And it's working okay?' I asked, making sure.

'Been ferrying stuff back and forth to 16 the past half-hour,' was what I got in response, and I took this to mean, 'Go for it'.

At just this moment, I happened to run into a buddy of mine, a strong, gutsy battalion chief named John Paolillo, and he let me know he didn't think it was such a good idea, me taking the elevator like that. He had his own orders, to head on up to a different high floor, to see about some different matter, and he actually tried to convince me to take the stairs with him, but I liked my way better. I had to get up to the 21st floor as quickly as possible, I had six men and tons of gear in tow, and those stair-wells were so crowded at these lower levels I feared it would take forever if we headed up on foot. And I liked my chances in this elevator. These maintenance guys had been up and down, up and down all morning, and there hadn't been a problem. But John was determined to take the stairs, so we went our different ways, and that was the last I ever saw of him. I'm not suggesting that my friend John Paolillo's decision to take the stairs wound up costing him his life, or that my decision to take the elevator wound up saving me mine and the lives of the men in my command, because the accounts of the next hour clearly suggest otherwise; but wherever John got to, however high he climbed and however he got there, he never made it back down, and this was where our paths last crossed.

Of the 99 elevator cars that serviced the north tower, only one stopped at every floor. This was the one firefighters were supposed to take for a garden-variety emergency. The rest were a combination of local and express lines, some running from the first floor to the 16th, and some reaching from 17 to 44, and some picking up on a high floor and running higher still. There were also several express elevators to the observation deck and to the Windows on the World restaurant, which opened on to various floors along the way. (The 'direct' ride, from the lobby to Windows on the World, was clocked at about 57 seconds.) Some of the shafts pulled double duty, carrying low-floor elevators, middle-floor elevators and high-floor elevators, and my only fear that morning was that one of the high-floor elevator cabs would break loose and come tumbling down on us as we made our way to 16. It wasn't a real, front-and-centre sort of fear, more like a concern, but once it landed in my thinking it was tough to chase it away. It wasn't enough to keep me grounded – frankly, I don't know if there was anything that could have kept me grounded at just that moment – but there it was.

I hurried my guys into the cab and pressed the button, and as the doors closed behind us I thought to myself, 'Okay, this is it. We're going. We're moving. No turning back. If something comes down on us, it comes down on us. Nothing we can do about it now.'

No one spoke on the short ride up. We were all facing dead ahead, lost in our own private thoughts of what would happen next. I imagine some of these guys were mumbling personal prayers of one kind or another, but not me; I wasn't much of a praying man before this day, so what came bouncing around in my head was more pep talk than prayer. I was pumping myself up, not calming myself down. I was the last one in the elevator, facing out, which meant I couldn't really see these guys from 110, except in what little reflection there was off the doors, so all I could go by was what I was hearing, and I wasn't hearing much beyond the whirr of the elevator itself. If it had been my own company, then maybe I would have turned my head and offered a reassuring look of some kind, or made some lame attempt at a joke, but as it was we stood silent. These guys were like brothers to each other, while I was a stranger to them. We were united in our effort, and in our purpose, but we had no history, and if it had been any other kind of situation, any *lesser* kind of situation, I suppose I would have made some kind of introduction. After all, if you're going into battle, you want to know who it is watching your back. But there was no time for introductions. I was Chief, and these guys were 110, and that was all we needed to know.

The precaution, when you're riding in an elevator in a compromised building, is to check the thing every ten floors, to make sure it's working properly, so I hit the buttons for the tenth floor and the 16th. You're also supposed to check it five floors below the destination floor, but I figured since we were just going to 16 I'd be okay bypassing regulations on this one. Ten, 11… 15, 16… what the hell was the difference? The doors opened on ten, no problem, and I didn't bother peeking out before they closed back up. None of us did. We just held still, and kept on with our staring, with our not saying anything. We were waiting on 16.

Sixteen was like a ghost town, like one of those scenes from a bad movie where some big trouble breezes in and clears out a room, and things are left in such a way that you can almost imagine all these people, reappearing back in place, picking up where they left off. Computer screens were still on, chairs were kicked away from desks as if their occupants were about to sit back down, half-sipped coffees and half-eaten muffins decorated almost every surface. There were loose papers everywhere, and all over the floor. Family photos smiled out into empty rooms. I didn't need to peek into more than a couple of offices to know what I would find in each: nothing but the recent whiff of some anonymous someone having been there, someone having bolted, without warning, someone not coming back anytime soon.

The elevator bank area was mostly empty too, save for the handful of firefighters there ahead of us. I motioned my guys into the nearest stairwell, C, and we began our ascent. There were still civilians coming down, but the crowd had thinned somewhat from what I had seen on the lobby level; what had once been a steady, overflowing stream was now closer to a trickle, and we weren't moving against the current so much as creating our own. It was now about an hour after the first plane had hit, so these stragglers were the slow-movers of the crowd, older people really dragging it, folks struggling in one way or another, stubborn mules refusing to give up their offices until the last possible moment. Still, there were enough people descending to create a stairwell traffic jam if we didn't take some proactive measures to guard against it, so I shouted up ahead to a fireman a floor or two above us. I could hear him bellowing directions: 'Stay to the left! Stay to the left!'

I thought I was hearing things. Stay to the left? I wondered what the hell traffic school this guy had gone to.

'Hey,' I yelled ahead. 'Stay to the right! Stay to the right!'

'I got you, Chief,' I got back. Then, 'You heard the Chief, people. Stay to the left! Stay to the left!'

We were one flight below this guy, and we were getting knocked around and short-circuited by all these well-meaning, misdirected people on the way down. We'd been climbing for just a minute or two, and already I was pissed. I took the steps two at a time and raced past the logjam, to see what this guy's problem was, and when I reached him I found a big, burly fireman, some guy I'd never seen before, doing his best to keep things moving. I grabbed his attention, and I told him again, 'Stay to the right! Stay to the right!' I even gestured with my right arm, to make sure he got it.

I still had no idea who this guy was, but he could see I was riled. 'I got you, Chief,' he said again, and then promptly spun around and faced the other way and continued telling people to stay to the left. I thought, 'Who is this idiot?' There was all kinds of confusion on this particular piece of stairway, people side-stepping each other, getting out of each other's way, hugging each other to keep from colliding. Tempers were about ready to burst – mine included.

I pushed up to him again. 'To the right, bro,' I screamed. 'The *right*!'

'I got it,' he still insisted. 'We're going up to the right, they're coming down on the left.'

He offered this last in explanation, only as soon as he said it he realized what an idiot he'd been. I just stared at him, didn't have to say another word, as this expression of stupid recognition washed over this poor guy's face. It's like you could see the light bulb go on over his head, and I turned him back around, and he went back to his business, telling people to stay to their right, finally.

It was a small crisis, and it strikes me as mostly funny now in the retelling, but it was maddening while it was going on. It unfolded in less than 30 seconds, but it was just about the last thing I needed, to have to deal with such incompetence, and I suppose the lesson here is that you can't expect people to think quickly on their feet, or even to tell their right from their left. Even us firefighters screw up sometimes – and some of us screw up more than others. If you do nothing, you'll never make a mistake, but you'll also never accomplish anything – and we firefighters accomplish a lot, even the few of us who occasionally screw up.

As we approached the 21st floor, I started to realize we might not be able to get there from where we were. For security reasons, not every floor was accessible from the stairwell; there were doors on every floor, but not every door could be opened from the stairs. For safety reasons, though, you had to have re-entry access every fourth floor, so as we made the turn on each full-floor landing I checked each door, to see if there was access. The doors opened 'in', towards the stairwell – that is, if they opened at all. There was no re-entry on 20, and no re-entry on 21, so I pushed up to 22, thinking I might find some interior access stairs, so I could double back down to 21. (Sometimes, if the same tenant occupies two or more adjacent floors, there are special stairwells cut into the design.) At the same time, my guys were banging on the doors on the way up, hoping there'd be some firemen or other rescue workers on the other side, which they were finding was often the case. We had our tools, and we could have forced the doors no problem, would have taken maybe 20 seconds, but you couldn't force every single door on the way up to the fire floor. Two guys, working a Halligan, which is sort of a combination of an axe and a crowbar, could have taken down virtually any door in no time flat, but even no time flat would start to add up, and we didn't have that kind of time.

By the time I doubled back to my guys from 110, they had already got to the floor, and we made a quick sweep of the area. I didn't stop to notice what kind of office it was, but they were all the same. Same layout, same drill. Computer terminals, hundreds per floor. Work stations. Scattered paperwork. Family photos. I'm the kind of guy who takes a mental picture of everything in sight, but there wasn't time to stamp a real fixed image. I saw everything, but very little of it registered. I was looking for people. Hurt people. Dangerous situations. Everything else was just background.

We conducted our search quickly and systematically. The design on most floors was built around a T-shaped corridor, with offices on the perimeter. We popped out of the C stairwell on the south-west corner of the tower, so I had the men splinter off in all directions, banging on doors, yelling. We didn't check every bathroom, every closet, but we made our presence known. 'Anybody here?' I hollered into the bullhorn. 'Anybody here?'

In less than a minute, we determined that the floor was clear, so we returned to the stairwell and continued up. Here again, we tried each access door along the way, and if it opened we stuck a little chock of some kind in the jam, to keep it from locking. Some of us carried a special plastic tool we used for just this purpose – a kind of sleeve that fits over the knob and latch mechanism – and we ran through our store of these pretty quick. There were already some firemen on 25, so the re-entry door was already open, and when I bounded into the corridor I told the lieutenant in charge that we had a report of people trapped here. These guys had been responding to the same report, and they'd already searched the entire floor, so I gathered the men in my command and considered our next move. It didn't trouble us in the least that we'd been dispatched to these floors, looking for hurt or trapped people, only to find that the floors had been cleared. That was the nature of the job, and the nature of an evacuation. Who knew how these distress calls had come in? Someone could have called their wife or their husband on a cell phone, describing the scene, and the spouse could then have called 911 and requested assistance. Or, a descending office worker could have noticed some trouble on his or her way down and made a report on reaching the lobby command station. Any number of things could have happened. What counted, though, was that the floors were clear, and that we had completed our assignment, and it now fell to me to figure where to go from here.

I looked at my guys from 110 and made a decision. I knew that in a high-rise fire, one of our operating tenets was to stay together as a company. I also knew that you can only go as fast as your slowest guy – and nothing against any of these tremendous firefighters, but I was carrying a lighter load, was in great aerobic shape and felt fairly certain I could make better time up to the fire floor travelling solo. In many ways, the department runs like the military: your company is your partner; you go in together, you come out together, you take your injured with you. That type of cohesiveness is terribly important in a raging fire – hell, in any type of dangerous situation – but I had to weigh this against the good time I could make continuing up on my own. If these guys had been from my battalion, if we had had any kind of history together, perhaps I would have felt differently, but my thinking, in those first few seconds once we realized we'd fulfilled our assignment on the 25th floor, was to move ahead without them. I could take these stairs like nothing at all, if I didn't have to worry about moving six other men with me. They had another 40 to 50 pounds of gear to carry, per man. Plus, there was a lieutenant among this 110 group, perfectly capable of leading his men and keeping the company together.

'Work your way up with your company,' I said to the lieutenant. 'I'll see you on the fire floor.'

The lieutenant nodded, as if this was the order he'd been expecting, and I took off. Right away, I started breezing past all these other firefighters, darting around them in the narrow stairwells like I was on a slalom course, taking the steps two at a time in some cases. I was a ball of energy, and I don't set this out to brag or blow smoke up my own butt, but to show how it was. The reason for it was obvious. These guys I was passing had been hoofing it since the lobby. I'd had a free ride to the 16th floor, so I was on relatively fresh legs. Put another 15, 20 floors below me, and I'd still be moving, but I wouldn't be moving nearly as well.

I hated like hell having to splinter off from 110, but I would have hated it more had I stayed. And I wasn't alone. There were a lot of singletons throughout the building, guys coming down off-duty, looking for their companies, or trying to hook up with another company. It wasn't a typical high-rise fire, in that there were dozens of firefighters on virtually every floor. There was always someone to watch your back, if it ever came to that, and in the split second I allowed myself to think things through I guess I figured I could do more good hurrying up to the fire floor than I could dragging the extra weight of an entire company.

Every few floors, on the way up, I stopped to check the re-entry door. Sometimes it opened, sometimes it didn't, and when it didn't I paused long enough to bang on it for a while, to see if there was anyone on the other side. Here and there, when I found an open door, I'd poke my head in to see if there was anything going on in the corridors or office areas, and then find some way to leave the door to the stairwell ajar for later re-entry. For the most part, the floors were abandoned. I'd find firefighters and other rescue workers, and from time to time I'd even see a civilian, but I contented myself in knowing that most of the office workers on these levels were already down. I knew full well that those poor people on the upper floors were suffering – indeed, it was now nearly an hour after the first crash, and those who weren't killed by the impact or the flames were at this point probably dead from smoke inhalation, but we all pressed on, thinking there'd be some people still to save.

I reached as high as the 35th floor – and really, it took just about no time to move those final ten floors. Less than a minute. When you're flying solo, like I was, travelling light, in relatively good shape, you can take those stairs like nobody's business. You feel like you could keep climbing forever. At the landing for the 35th floor, though, I noticed the re-entry door had been propped open, so I stepped into the corridor area and went over by the elevator banks, where there was a good deal of activity. Some-how, this had become a kind of staging area, with maybe three dozen rescue workers waiting to be deployed. Firefighters, mostly. There were a couple of police officers, and a Port Authority cop or two. Some had reached to higher floors and been sent here for reassignment; some had parked here on the way up, and were waiting on orders. I spotted a couple of familiar faces, asked a couple of guys how they were doing, what they had found, what they were hearing. I didn't want to waste too much time here, but I wanted to get the lay of the land.

And then it happened. That deafening, sickening noise from above, like nothing any of us had ever heard. Like an earthquake mixed with a thundering herd and a fleet of runaway trains. All at once. And all headed right at us.

FOUR: Retreat

People a couple of miles away could hear this noise, and they described it as 'monstrous', 'deafening' and 'thunderous', so you can imagine what it sounded like from on high, just across the plaza. I didn't even want to think how it was from its middle, for those poor souls trapped inside at the moment of collapse, but there was that thought, too.

It took only ten seconds for the south tower to come down, which by my reckoning meant that if we were on the 35th floor it took about six and a half seconds for the rumble to reach us, a beat to pass through us, and another three and a half seconds to rush away from us to the ground below. It might seem like splitting hairs, to break it down in just this way, but it was important to me to understand the timeline, to reassemble those moments. I felt I needed to know how we were managing with each tick of the clock, what each second meant, and what each second cost. I felt this way at the time, and I feel it now in retrospect. Who knows why we grab at the small details of our existence, or why we hold on to the oddest pieces of information as if they carry some clue? I can't even guess at it, in my case, but the numbers corresponding to this stretch of time struck me as key. Certainly, they were interesting, and I collect them now as pieces of evidence which, together, might add up to something. As it happened, it seemed longer than ten seconds. Much longer. 30, 40 seconds, maybe, and that's a good long time, to be terrified over something you can't really know. And in those ten seconds, there was enough time to think through every eventuality, every cause and effect, every damn thing there was to consider. That's the giant irony of a tense moment like this one, the way it gives you all kinds of time to think things through and no time at all to move a muscle. You've got all the time in the world, and no time at all, both.

Behaviour scientist-types have a name for this moment of indecision, this brief inability to decide on a next move at a point of crisis. It's called the 'fight-or-flight reflex', and it's something we firefighters are made to study, as we learn to harness our actions and reactions in emergency situations. Or, at least, to understand them. It's a reflex as old as man. You're running, or going about your business, and all of a sudden a bear or a sabre-toothed tiger turns up in your path. Our natural survival instinct tells us we now have two options: we can stand and fight the beast, or we can turn tail and run, and what's fascinating to me is that we land on one or the other without even thinking about it. It's instantaneous, and pure instinct, and it's all tied up with issues of personality and perspective. Until that moment, you can't know which way you'll go. You make an immediate assessment of your chances. You're looking at the tiger, you're wondering if you can make it safely to that nearby cave, or maybe you like your chances better if you stand and fight. I suppose there's also a third option – to simply roll up and die – but we firefighters tend to dismiss this one as absurd; it might be human nature for some folks willingly and passively to accept their fate, but it's not in *our* nature, so it doesn't even get discussed.

And so those ten seconds passed, slowly, and with them the bone-chilling roar and rumble, and the fear that some awful fate would find us in this building, and after a while the tower stopped shaking and we started thinking in proactive terms. We were all still okay; we were all still standing. Our fight-or-flight reflexes kicked in, only here it took a few beats more to understand the situation. It was more complicated than a sabre-toothed tiger in our path. The noise and the shaking stopped, but we still had no idea what kind of danger had just presented itself, or what it might mean. The commotion could have been anything, and our first order of business was to assemble the possibilities. We got the report on the tactical channel, that the south tower had come down, and that seemed to root our thoughts along certain dire lines, but there was no one on the command channel issuing any kind of directive or explanation or description, no one to be our eyes and ears on the outside, so we slowly gathered our wits and began to think ahead. It's amazing to me now, in retrospect, that no one among us thought to race to the south-facing offices to look outside, to see for ourselves what had just happened. It's amazing to everyone who hears my story. And yet what I can't adequately describe, or even understand myself, is the way that such a move didn't even present itself as a prospect. Yes, we were just a quick dash away from a mess of windows that might have lent some perspective on things, but I guess it didn't seem important at the time. We had more pressing matters to consider. And besides, with all the smoke and ash and concrete dust that had clouded where the south tower used to stand, we wouldn't have seen anything had we looked out of those windows, beyond smoke and ash and concrete dust.

Gradually, I started to notice that everyone's attention was shifting towards me, looking for our next move. I was a battalion chief, it said so right on my bunker gear, and there was no one else around with any higher authority. We'd all been conditioned to follow the chain of command, and these men were looking to me to make a decision; their instincts were to trust mine, and there was no time to base my actions on anything but gut. There was no one on the radio, no hard information, just the simple, daunting fact that the south tower of the World Trade Center had collapsed, and the all-but certainty that whatever had brought that building down was going to bring this one down also. And soon. A bomb. A secondary explosion of some kind. Another plane. Whatever it was, we were vulnerable. More than we had been just a few minutes before. More than we had imagined. This was my thinking, at just that moment, and I added to this the probability that the only people left alive in the north tower were rescue workers, just like this group of firefighters at my side. At this point it was over 70 minutes since that first plane had hit, so I had to assume that even the slowest-moving office worker had by now descended below the 35th floor, where we now stood. I couldn't imagine there was anyone still alive above the fire floor, with the way the fire had been raging and with all that killing smoke. And I couldn't think how anyone might have managed to get from above the fire floor to below without assistance. The predominant life hazard, here on in, was firemen, policemen and other emergency workers, no doubt about it, and the more I thought on this, the more I realized I had to get these people out. Me. Rich Picciotto, Battalion One One. There was no one else to make the call.

With the collapse of the south tower, I went from being a lone wolf chief, racing up to the fire floor to lend a hand, to the highest-ranking officer in the immediate area. Just like that. As far as I knew, I was now the highest-ranking officer on the highest floor of the tallest remaining building in New York City. No one was higher, which meant that I was suddenly cast in a completely different role. Remember that John Mellencamp song? 'When I fight authority, authority always wins.' Story of my life, but I always kept fighting, and here I was, authority itself. With no one or nothing to fight, but this impossible situation.

I've got to tell you, this was not an easy call to make, and I hated like hell that it fell to me. Actually, let me rephrase the front end of that statement. There was no other call to make, but it was hard to be the one to have to make it. I tried desperately, frantically to get on the command channel, to find someone else of higher rank to assess the situation, but I got no response. I wasn't looking to pass the buck so much as to follow protocol. Throughout the department, there were chiefs of higher rank, but even I could see no one was in a better, more front-line position to call for an evacuation. There were about 50 deputy chiefs and ten staff chiefs in the department, and there was no way to know how many were on the scene. Plus, even if they were in the vicinity, who the hell knew where they were in relation to this building, or how many of them were still alive now that the south tower had collapsed?

Naturally, any firefighter could call for an evacuation, but there's a certain comfort in knowing the move has been validated by the brass, that it flowed from the top. To change our course of action, from rescue to retreat – that's a huge shift. It's huge in the abstract, and it's huge in the execution. You go from having a couple of hundred guys moving up, higher and higher towards the fire floor, to having a couple of hundred guys moving down. It's like a massive, shifting wave; all along the tide is going one way, and then it suddenly goes another. Fight or flight, that's what I was thinking. We couldn't run from a situation like this, not from the 35th floor, but at the same time there was nothing left to fight. It was unlikely that there were any more lives to be saved, other than our own, and if I was wrong on this I knew we sometimes had to think about sacrificing the few to save the many. Clearly – or, I should say, most probably – this was one of those times.

So I did what I had to do. I brought my radio to my lips and called for the evacuation of the north tower. It was probably 10:01 am, give or take a couple of ticks. I wasn't looking at my watch, but it couldn't have been more than a full minute after the collapse.

'Get out!' I hollered. 'Let's start moving! Drop your masks! Drop your tools! Drop everything! This is an evacuation! Let's move it out!' I darted from the bank of elevators to each of the three stairwells, and yelled up and down in each with my bullhorn, 'Get out! Get out! This is the fire department! We're evacuating! Get out, get out, get out!' I put it out on the command channel, too, but like I said, our radios weren't working too well in this particular piece of air space, so I had no idea who was hearing me on the other end. The guys in my immediate view, though, they heard me just fine, and they were dropping their gear and bugging out. It was an orderly retreat, but it was very definitely a retreat. In the department, we have an expression, 'assholes and elbows'. Picture someone running from a burning building and what do you see from the rear? Assholes and elbows, right? Well, that's the phrase we use to describe a hasty, frantic retreat, and this was no assholes-and-elbows evacuation. This was a collection of highly trained, even-tempered professional rescue workers, following orders. There was no mad rush. If anything, there were a couple of guys wanting to be a little more prudent with their equipment, thinking they could get down just as well with their gear as without, but I impressed on these guys the importance of haste in this particular situation. I told them, if the building didn't come down, we could always retrieve the masks and the tools and the extra cylinders at some later point; if it did come down, we certainly didn't need all that extra weight slowing our retreat.

Even now, in the retelling, there's something about that word that bugs the hell out of me. *Retreat*. It's not normally in my vocabulary. When it comes to that fight-or-flight reflex I talked about earlier, I'm fight. All the way, all down the line, all the time. No doubt about it. But here I felt I had a responsibility that was bigger than just me. Whatever small chance remained that we would find someone alive on one of the upper floors, I had to weigh that against the far bigger chance that the few hundred of us on these lower floors wouldn't make it out of the building. There was no choice but to surrender, to give up the building in order to save our own hides. Don't get me wrong, I wasn't second-guessing myself, but there was a part of me felt like I was abandoning someone, somewhere on one of those 110 floors. That's how I looked at it, like I was giving up. I'd never knowingly left anyone behind in a burning building before, and here I'd based my call for evacuation on the simple implausibility of finding anyone alive – at this late hour, over such a vast piece of real estate. It'd be like looking for a needle in a burning haystack, I knew, and yet if it had just been me, I probably would have gone looking.

But it wasn't just me. It was hundreds of firefighters and police officers and Port Authority rescue workers and emergency medical technicians, and it was all their families on the ground, praying for their safe return. It was the whole world watching. Really, it was a big bunch of stuff and circumstances and hopes all rolled into one big picture. So I lifted the bullhorn and sounded our retreat, up and down the stairwells and all along the corridors. I also gave the order on both channels. I actually pulled a handi-talkie off the chest of a nearby fireman, to put it out on the tactical channel, which seemed to me more efficient than changing the channel on my own radio, and then I barked into my own chest for the command channel order. And nobody questioned it. Nobody said, 'Hey, Chief, we still got work to do here,' or anything like that. They simply dropped their gear, and turned, and began to descend.

Finally, about a minute after I called for the evacuation, I heard an authoritative voice coming through the static on the command channel. 'Who gave the order to evacuate?'

I thought, 'Oh, shit!' I thought, 'Man, I just blew it!' I thought my career was over. Whoever it was sounded real pissed, and I reminded myself what a big deal it was, to call for an evacuation on a job of this magnitude. Biggest decision I ever had to make, in 28 years in the department, and a minute later I was being called on it.

I keyed the mike on my radio and answered back. 'Battalion One One to command post,' I shouted. 'This is Chief Picciotto, Battalion One One. I gave the order to evacuate.'

I didn't get anything in response, which I took to mean whoever it was questioning my call hadn't heard me. Sometimes, in a high-rise, with our piece-of-shit radios, there are dead spots transmitting but not receiving. Sometimes you can talk but not hear. And sometimes the damn things don't work at all. I wanted to own up to my call, to take any consequences. I'd made a decision. I thought it was the right decision at the time, and I know it was the right decision now. But, that said, it still felt like I was being called in to the principal's office, the way this lone voice came over the radio, wondering what the hell was going on.

I should mention here that I didn't always get along with my superiors in the department. I had a reputation for butting heads with the administration, for making the kind of call that didn't always consider the bottom line or the public perception, so I was conditioned to this kind of dressing-down. Usually it was over stupid stuff, but I never shrank from a fight. I was constantly standing up for my men, railing against this or that piece of unfair treatment or foolishness. Lately, the head-butting seemed to be about the trimming of our budget and the cutting back of our manpower and the prehistoric nature of our equipment. The constant cost-cutting by our current administration left us firefighters on the line feeling our administrators were putting pennies ahead of lives, and I was never one to keep quiet about it. If I saw bullshit, I called attention to it, and I made sure everyone else knew it was bullshit. I could be a good soldier, at times, but I could also be a thorn, which was why, when I put my name out there, when I stood behind my call, a part of me felt like I was about to be nailed. When I fight authority, authority always wins, right? These guys upstairs were always looking to bust me down, transfer me out of the borough, screw me in some way; it's like I had a bull's-eye on my back and here I was an easy mark.

Still, there was no response, so we continued with our evacuation. A few beats later, though, I heard a second voice on the command channel telling us to stand by. For about 15 seconds, we all stood still – that is, all of us in my view, still on the 35th floor, which at this point amounted to about 20 firefighters – although I imagine the scene was much the same on every other floor, where the radios were working. It was like a weird game of administrative freeze tag, the way they put us all on pause like that, but before I could get too bent about it I heard a staff chief come on and instruct us to continue with the evacuation. Whatever bullet I thought I'd been dodging had shot past.

Very quickly, I established a routine as I cleared

each floor, beginning with the 35th. I was closest to the C stairwell when I started my sweep, so that became the route for my descent. The C stairwell, as I have described, was on the south-west corner of the building; the A stairwell was on the south-east corner; and the B was at the midpoint or core. On most floors, the layout was much the same, as I have also already described, but I will do so again here, to keep things straight. The corridors were fashioned along a 'T', with the A and C stairwells at the wing-ends of the T, and the B stairwell at the butt end. The interior offices fed off these main corridors, and depending on the way each floor was leased, to one tenant or several, the floor designs ranged from open to closed. By open, I mean it was possible to see inside certain offices, through glass walls and such, and by closed I mean it was like the stark hallways of most office buildings, where you couldn't tell one floor from the next but for the numbers on the walls. On some floors, the layout was so open there was no noticeable hallway. So it varied, even though the drill was the same.

The drill was for me to start at the C stairwell and race across the south-facing hallway to the A stair-well, banging on office doors along the way, hollering for everyone to get out. At the A stairwell, I'd step inside, hop a couple of steps up and shout into my bullhorn, skip a couple of steps down and shout again, and then return to the re-entry door. From here, I'd double back along the south corridor to its midpoint, where I'd veer right for the B stairwell, once again banging on doors and shouting out instructions. Next, I'd clear the B stairs like I'd cleared the A stairs, and then I'd double back again to the C stairwell, where I'd started out. In this way, I cleared each floor, bringing up the rear in our evacuation. It sounds like a lot of effort, in the recounting, but it was possible to cover the entire floor in about 40 to 50 seconds, and by the time I'd reached the C stairwell there'd still be a logjam of rescue workers making their way to the next landing. The going was so slow in there, and the stairwells so narrow, especially for a bunch of burly, turnout-coated firefighters, that I really didn't lose any time sweeping each floor in this way. I'd hurry up and wait, hurry up and wait, one floor to the next.

I didn't abandon the 35th floor straightaway, either. There had been a good number of people scattered on the floors above, so I waited for them to pass before I could determine that the upper floors had been cleared. When the flow of firefighters and other rescue workers slowed on 35, that's when I began my descent. As long as they were in the stairwells, I felt it was okay for me to push down. I figured that the ones starting out higher than me would soon enough leapfrog past, as I ducked out on each floor to do a sweep.

We moved without incident until about the 31st floor. I can only approximate the location here because on most floors there was little to distinguish one from the next, but we had cleared about four more floors when I came across a small snag adjacent to the B stairwell. There was a cluster of firefighters and Port Authority police officers, and the focus of their attention was a middle-aged, Middle Eastern man. The guy was dressed in a nice suit, and carrying a nice briefcase, and as I pulled close to the commotion I could see there had been something of a struggle.

'What the hell's going on?' I barked into the hallway. I didn't need the bullhorn to express my contempt that my people weren't moving swiftly down these stairwells. I could be a bit of a hard-ass on the job, if I felt my orders weren't being followed, or if I felt we were putting ourselves at unnecessary risk, and this was one of those times. I didn't like snags, or surprises, and I made my displeasure known.

'Got it under control, Chief,' one of the firefighters responded. 'They've taken him into custody.' With a nod of his head, he indicated the two Port Authority cops at the centre of the small scuffle. What had happened, I quickly pieced together, was that one of the firemen on the floor had grown concerned at this man's sudden appearance; the firefighter had once been a police detective, and he felt there was something fishy about this guy, something about his clothes or his briefcase, something about his still being on such a high floor so long after the first crash. By this point, all kinds of speculation had been bouncing around on our radios regarding responsibility for these terrorist attacks, and the sight of a transparently Middle Eastern individual was suspicious. Just as likely, this well-dressed office worker stood out not for any reasons of racial profiling, but because he was out of uniform. Plain and simple, everyone else on these floors was an official rescue worker, and the sight of this civilian set off a small wave of worry. However it happened, this detective-turned-firefighter pointed this guy out to two Port Authority police officers, and together they figured that if this individual was still on the scene, still on one of these upper floors, he could only be up to no good. Plus, he had a briefcase, which could have contained a bomb, or some other device aimed at doing us dirt.

By the time I happened by, making my sweep of the floor, the 'suspect' was in handcuffs, and crying uncontrollably. He was claiming his innocence of any wrongdoing – as if a true terrorist would not have done the same! – but it seemed to me the tears were for the sudden fear that he would die in a building collapse. I thought, 'Shit, if the guy's so afraid he's gonna die, what the hell's he doing dragging his feet and taking his sweet time getting down?' It was too long after the first plane hit for someone to be panicked over his circumstance.

Nevertheless, it wouldn't do to have this incident bog down our evacuation, and it wouldn't do to have this guy crying his eyes out for the next 31 floors, so I moved the whole group of them along as best I could. I figured that if the cops wanted to cuff him, they could cuff him, but they all needed to get out of here just the same. His guilt or innocence could be established later. Or, not. And as I watched these Port Authority police officers lead this well-dressed office worker away, his hands bound behind his back, I allowed myself to marvel how odd it was, to be facing such massive, incomprehensible disaster and at the same time dealing with such petty nuisances. In all likelihood, this guy was guilty of nothing more than foolishness and slow-footedness, but here we were, escorting him down as he cried chickenshit tears, and as we fought back our own terrors of what was to come – here, in this building, and out there, in the rest of the world. I shook my head in wonder; it was barely an hour since the first attack and already we were running scared.

We moved another few floors without a hitch, to about the 29th floor, where I reached another staging area in one of the hallways. Actually, it wasn't so much a staging area as it was a gathering place. A rest spot, really. Just a bunch of guys, maybe a dozen or so, stepped out of the stairwells for a breather, was how it seemed to me at the time. I knew some of these guys, by name or by face or by reputation, so I did what I could to hustle them along. One of them was waiting on another guy from his company, but I assured him that the guy was already down. I figured the white lie wouldn't hurt anybody, and I was afraid that without it we were all at risk. See, I'd set it up in my head that I'd be bringing up the rear on this evacuation – I guess you could say I was standing behind my own call, in a very literal sense – so for me to allow this firefighter to wait on his partner meant I'd have to wait as well. I wasn't leaving anybody behind, so I told him I had it on the radio that this guy had been spotted on a lower floor.

Even with a white lie here or there, I had a tough time moving some of these men along. They all had people they were waiting on – chiefs, lieutenants, brother firefighters – and they weren't too happy about moving down without them. There were police officers, too, waiting on their partners, and at least one Port Authority cop waiting on his chief. I heard later that one Police Department boss actually sent his men down from this point, while he himself hung back, as if he was waiting on another of his men, only instead of waiting he ducked into one of the offices, fired up a cigarette and kicked back. The story that found me some time later, and I can't say for sure whether or not it was true, was that this guy had been having all kinds of personal problems leading up to this day. Health problems, money problems, marital problems... the whole deal. Apparently, from what I heard, it was all too much for this poor fellow, and I guess he figured his time was up anyway, so he closed himself inside an abandoned office and waited to go out on his own terms. Again, I've got no idea if this really happened, or if it was just one of those legends that reached us on the way down, but I include it here for the image, for the way it reinforces what was racing through all our minds, in one way or another, as we raced to get out of there. There wasn't a man among us who wanted to be seen in any kind of assholes-and-elbows rush for the exits, and I'm guessing there were quite a few who would have found some dignity in standing firm and pressing ahead to the fire floor, even against the long odds of making any kind of difference. And there are always a few in every bunch having various troubles in dealing with their personal lives. Mix those truths together and you're bound to have at least a few folks sitting this one out, waiting for their fates to find them on one of those upper floors. Kinda like those musicians on the deck of the *Titanic*, playing away while the ship went down, knowing there was nothing to do but stay resigned to their fates with a song in their hearts. So I set it out here, for the way it gets at the various moods that found us on the way down. I've got no hard evidence to believe the story, but I've got no reason to doubt it, either.

It was then that I ran into the last able-bodied civilian I would see on the descent – a well-dressed broker-type hunched over his desk, typing crazily on his keyboard. I noticed him on one of my sweeps of the perimeter office areas, and at first I thought I was seeing things. I was on or about the 27th floor, bringing up the rear, and along a row of computer terminals there was this one guy. It wouldn't be accurate to suggest he was working as if he had no idea what was going on, because clearly he knew the situation, but there he was, banging away on the keys. It was such an unexpected sight, but frankly I was more pissed than surprised. This was an evacuation. We were the fire department. I had issued an order. And, in an emergency situation such as this, I expected that order to be followed. Fireman, cop, civilian, mayor, President... I didn't care who you were, I expected you to listen. So I hollered over to this guy. 'Hey, what the hell you doing?' I said. 'This is the fire department! We're getting out of here! This is an evacuation!' I wasn't nice about it. I wasn't mean. I wasn't anything. I was just hollering.

At this, this nut simply held out his hand, palm-up, in a stiff-arm, like he was a friggin' traffic cop. 'Wait a minute, buddy,' he said, all entitled. 'I got something important here.'

I was stunned. During a fire, no one tells me to wait. This was a life-threatening situation. His life. My life. Lives all around. He might have felt he was free to do as he pleased, that someone like me had no authority over someone like him, but I wasn't leaving anyone behind, and I sure as shit wasn't waiting on this guy's well-dressed ass. He was about 35 years old, white, and he seemed completely unruffled by the events of the past hour. From my angle, it was impossible to tell what it was this guy was doing that was so important. I'm guessing now he was downloading some information, or maybe backing something up on to a disk he planned to take with him. I don't think he was making any kind of trade or anything like that. By this point, the phone lines were mostly down, and this didn't appear to be a trading floor, but ultimately it didn't much matter what this guy was doing. What he was doing was ignoring my order, and putting me and my men at risk.

'This is the fire department, buddy,' I shouted again, moving towards him now. 'This is an evacuation.'

This time, he just gave me the stiff-arm, didn't even make the courtesy of eye contact, and I saw this guy's open palm and just flipped. 'Get out!' I yelled, as I grabbed him by the shirt. 'Get out!'

I grabbed him with both arms, lifted him from his chair, and hoisted him over to one of the firemen who had followed me on to the floor. 'Get him the fuck out of here!' I told the firefighter, who in turn shoved the man along towards the exit.

It was the strangest, most frustrating scene, and what got me most of all was the way this guy looked back at me in disbelief. Like I'd crossed some line. Like it was me who'd gone completely crazy, not him. I'd ripped his shirt in the heaving – a real nice silk shirt – and he kinda stumbled to the stairs, shaking his head, incredulous that some sweaty civil servant had bossed him around and manhandled him in just this way. And as he disappeared into the B stairwell, I thought to myself, 'Gee, I feel real bad about having to rough this guy up, to rip his shirt like that.' And then I thought, 'Yeah, right.'

For the most part, though, it was an orderly evacuation, with no real surprises beyond how smoothly things seemed to be going. At this point, at about ten-fifteen or so, about an hour and a half after the first plane struck, there were no longer any office workers in view, other than the well-dressed asshole on 27 and the handcuffed 'suspect' on 31. Everyone who was getting down had got down. Every once in a while, though, I'd run into a firefighter looking for some guy from his company, or his superior officer, and here again I'd reach back for one of those white lies I fell into telling on some of the higher floors. I was caught between being a good guy and being the chief, and I couldn't be both. I'd met someone on 25, say, who'd last seen his lieutenant on his way to 28, so I'd ask for the lieutenant's name, or for a description, and then I'd say with absolute conviction, 'Oh, yeah, he went down one of the other stairwells. I was just talking to him.' Or, whatever it took, in each case, to get each guy moving. They'd ask me if I was sure, and I'd say, 'Yeah, I'm sure.' In truth, I'd have no idea, but I would have said anything, because I wasn't leaving a single person above me.

People have since asked me if I had any trouble with my actions in this regard, if there was any kind of moral dilemma keeping me up at nights as a result of these not-quite-half truths. And, honestly, I don't see that I had any choice to do any different. I don't see one now, and I didn't see one then. A part of me thought that in a high-rise fire, companies are supposed to stay together. That's rule one. Even the greenest proby – or, probationary firefighter – knows that. So, if a company is not together, it means someone screwed up, somewhere along the line, and I didn't see that I should put myself in jeopardy, or any of these other men who were doing their jobs and following orders. That said, I also realized that shit happens. You're only as strong as your weakest link, right? Maybe someone's knee starts acting up, or his back goes out on him, from all that extra weight he's carrying. It sometimes happens that a company has to ask one of its guys to hang back, to wait on their return, because he's slowing them down on the way up. We've all been in those kinds of situations. I didn't mean to be hard about it, but the truth of it was I didn't even think of these white lies as lies at the time. I'd given an order, and I expected it to be followed. Whatever the situation. Whatever the special circumstances. You start telling me reasons why you shouldn't have to follow my order, you start making me explain or defend my decision to evacuate, or try getting me to change my order, then I have a problem with you. This was an emergency situation. I didn't have the time to argue the point with 400 different firefighters. That's about how many firefighters there were still in the building. Of course, it didn't happen 400 times, but it happened three or four times, and that was three or four times too many.

Of course, if some superior officer had told me to evacuate, and I knew for a fact that one of my men was a couple of floors above me, in some kind of distress, I don't know that I would have gone down dutifully. If the situations had been reversed, I'd have been more thorn than good soldier. No question. I'd have probably ducked into one of the offices, out of view of the battalion commander ordering the evacuation, and waited until he had cleared the floor before moving up to find my guy.

FIVE: Stall

We fell into a routine, a rhythm: clear the C stairwell, cross to the A, double back to the B, return to the C, descend, repeat. Less than a minute on each floor, a few seconds more if I needed to do some coaxing along of left-behind men. Nothing to distinguish each floor but the metal markers on the stairwell walls, the thinning crowd and the steady progress we were making.

In all, it was a fairly smooth operation, made swifter and surer by the professionalism of most everyone involved, and with each floor cleared I started to think we'd make it down without incident. Everyone beneath the fire floor, anyway. Everyone we had a shot at saving. Every rescue worker who'd raced into the building shortly before nine o'clock that morning. What had seemed a long shot while we stood on 35 now seemed increasingly likely, as we reached below the 20th floor and into the teens. At this rate, in these conditions, we'd be down from here, and out, in no time.

It's important here to emphasize that I wasn't the sort of firefighter who struggled with thoughts of gloom and doom. They were a constant, a part of every job, but I never let these thoughts get in the way or slow me down. I once heard someone describe it as a little bit like the way you come to grips with the concept of life and death as a child; you learn that death is a natural part of life, and you learn to deal with it. It's the same on the job; death is a part of what we do, and we learn to deal with it, and once I learned to deal with it I never worried too much about any of the bad shit that could find me on any one of these jobs. To tell the truth, I don't know too many firefighters who spend too much time worrying. Planning, yes. Considering every eventuality, yes. But worrying? Nah. What the hell's the point? It's always there, like I said, back of your mind, and we're always careful to guard against every conceivable situation, to take every precaution; but we set our worries aside, best we can. With me, out on a job, I rarely thought about Debbie, or the kids, unless it was in a very specific way, for a very specific reason. If Stephen had a wrestling match in the afternoon I meant to get to, for example, I might look at my watch and wonder if we'd be through with a job in time for me to make it, but other than that I was focused. The rest of the world fell away. Even thoughts of friends, and family. And there was never a thought for my own safety. Always, I'd do what needed to be done, whatever needed to be done. If I made it out, I made it out; if I didn't, I didn't, but God knows I would die trying.

Since 11 September 2001, people have asked me when it was that I first realized I might die in the north tower of the World Trade Center, and my answer always sneaks up on them. I tell them the thought first found me in the drive downtown in the rig. Not because of the breakneck way Gary Sheridan was driving, but because of the dawning realization of what we were driving towards. Always, when I'm heading out to a job, it occurs to me that it might be my last job. It occurs to me, and then it disappears. Big jobs, small jobs, and in between. It's the nature of what we do. Anyway, it's my nature; can't say for sure how it happens for anyone else. Me, I have this thought, and then I let it go, and it doesn't come back until the next job. At least it never had. It would come back to me again on this day, this confrontation with my own mortality, in ways I couldn't possibly imagine, but for the time being it was the furthest thing from my mind. Here I was, racing past each floor – 20, 19, 18 – and all I was thinking, really, was that we were making good time.

And then, on or about the 17th floor, our progress halted. We were moving down the C stairwell, when we suddenly ran out of room. Everything stopped, and everyone bunched together, and the narrow stairs felt like a subway platform at rush hour when the trains weren't running. We were backed up good, with no place to move. And it wasn't like we all slowed. We stopped. Dead. All of us. All at once. All down the line, like we hit a brick wall – or, as it turned out, a wall of ruin.

Remember, this was not a wide stairwell to begin with. 42 to 48 inches, return-style stairs. Not a whole lot of room to work with, in the best of circumstances, and these were far from the best of circumstances. We'd been travelling one or two abreast, and moving freely, but here we were made to stop. There was something down below that had clogged the way, and none of us on these higher floors could figure out what it was. Remember, too, that these were trained, composed rescue workers, firefighters mostly, so when we got all bunched up like that there was no immediate sense of panic. We'd been through such as this before. There was some quiet rumbling, some general figuring out of what was going on down below, but there was no disorder or confusion. In fact, I didn't even hear any commotion until I ran smack into it. I'd done my hurry-up-and-wait routine on 17, and hopped back into the C stairwell, and half-way down to 16 I noticed the jam. I got on the bullhorn right away, shouting down into the space between the interior railings. 'What's going on?' I tried. 'Let's keep it moving! What's going on?'

As I spoke, bullhorn pointed down, I could make out between the railings that we were jammed up as far as I could see. 'What the hell's going on?' I tried again. 'What's slowing us down?'

Soon enough, it filtered back up to us that the stairwell had become clogged with debris. As far as we could determine, when the south tower had come down, all kinds of glass and concrete and rubble had filled the mezzanine of the north tower, and possibly reached up into the open stairwells as well. How it was that we hadn't noticed the jam before, I couldn't imagine. There are all kinds of explanations. It's possible there had been a narrow opening, big enough for people to slither through, one by one, but there had recently been some sort of secondary collapse or shift in the rubble pile that now made it impassable. Who could tell, from where we stood? Someone could have been rerouting traffic on a lower floor, and once there was no longer any strong flow of civilians coming down those stairs they might have moved on to some other task, which might have left the somewhat-later-to-arrive rescue workers to discover the impasse for themselves. It could have been anything. All we knew was that we had been moving, and now we were not moving, so I immediately clambered back up from the landing midway between 16 and 17, and raced through the corridor on 17 to the A stairwell. But the situation there was much the same. This made sense. Both stairwells fed out on to the mezzanine level on the side of the building facing the south tower, so if there had been enough debris to choke one passageway it was likely it would have choked the other. Next, I shot over to the B stairwell and noticed it was thankfully clear, so I raced back and redirected the traffic jams from A and C to the interior stairs.

Here again, it was an orderly shift. No mad rush. No hot tempers. I never determined how far down people had been when traffic stopped, how high the debris had reached into the south-facing stairwells, but I managed to direct people to the interior stairs on each floor – by bullhorn, by radio, by a kind of vocal bucket brigade, where we passed the word on down the line, one firefighter to the next, that the B stairwell was clear.

And so we moved. I continued with my sweep on each floor, and more and more I was coming up empty. The people who had been backed up in the stairwells had found their way to the interior stairs, and the floors were mostly cleared. On the 12th floor, though, I felt for a moment like I had stepped inside an old 'Twilight Zone' episode. Truly, it was the most bizarre thing. I raced from the B stairwell to the two south corners, and on this floor there was an odd configuration of office space. To the south and west, next to the C stairwell, there was a glass-walled office area, while the rest of the corridor appeared to be of the standard, no-frills variety. The glass was tinted in such a way that I couldn't see clearly through to the other side, so I opened the door and made to shout out my evacuation orders. Once inside, though, I caught the strangest sight. Forget the 'Twilight Zone' comparison; it was more like stepping inside one of those 'What's Wrong with this Picture?' drawings, from those old children's magazines. And what was wrong with this picture was this: the office was filled with people. Forty, fifty, sixty of them. I didn't stop to count, but there they were, in all shapes and sizes, of all ages. And all of them were sitting quietly, patiently, apparently waiting for instruction or assistance.

I thought to myself, 'Man, what the hell happened here?'

As I stepped in, I noticed another firefighter on his way out. 'Chief,' he said, intercepting me, 'we got a problem.'

I thought, 'No shit.' Most of these people were sitting. Some were standing, but most were sitting – many of them behind desks or partitions and placed in such a way that I couldn't get a real good look at any one of them. In my haste, and surprise, I simply took in the whole lot of them, all at once, and raced to my own conclusions. I thought I saw the scene for what it was, and I didn't give the fireman a chance to explain himself fully. If I had, I might have saved myself the trouble of putting two and two together, which in my head didn't quite add up.

'That's all right,' I said. 'Let's get these people out of here.' And, without waiting for any kind of response or acknowledgement, I started barking out orders to this group of wide-eyed office workers. I couldn't make out any panic or urgency in the sea of faces looking back at me. Everyone was outwardly calm. Naturally, this group demeanour didn't fit with what was happening elsewhere on this plaza, on this morning, but I was so busy trying to move people down that I didn't stop to notice. If someone had told me to stop and consider what this collection of people was doing, congregating in this interior office, waiting to be told what to do, I wouldn't have had the first clue. I didn't have time to play detective, but as the scene registered for what it was, I started wondering. From the looks of things, it didn't appear to be a group of people from the same office, or even from the same company. They were wearing assorted styles of clothing – some casual, some more formal, and so on. Plus, they didn't seem to be sitting at their own personal desks or work stations. Evidently, these people had been parked here for some purpose. What seemed clear to me, in a quick leap of logic, was that this wasn't their office, and they weren't all colleagues. They were assembled here for some unknown reason – and whatever it was, it struck me as no longer relevant. The time had come to move them out. And down.

There happened to be a large group of rescue workers congregating in the corridor just outside the glass doors to this office, moving from the clogged south-facing stairwells to the relatively open passage of the B stairwell, so the hallways were fairly choked with burly, grimy, helpful men. I shouted out the office doorway for the men to clear a path. 'Office workers coming through!' I yelled. 'We've got civilians in here! Clear a path!' And on my orders, these great, good men fell to the side in such a way that it almost looked choreographed. Almost immediately, all down the north–south hallway to the butt-end of the T-shaped corridor, and on into the B stairwell itself, this sea of mostly firefighters parted to create a path for these inexplicably trapped civilians. To a man, these rescue workers were anxious to get down to safety, but these office workers would go first. And, also to a man, they appeared grateful finally to have a job to do. For too many of them, this presented the first piece of rescue work they'd encountered all morning, and they fell into it with enthusiasm.

Now, here's where the really unusual piece kicked in. Once the path had cleared, these office workers started moving towards me. I got on the bullhorn, wanting to make sure they heard me above the din in the room. 'Pay attention, everyone,' I said. 'This is Chief Picciotto of the 11th Battalion, New York City Fire Department. This is an evacuation. Move as quickly as possible to the B stairwell at the centre of the building. There are firemen in the hallways to point the way. Move as quickly as possible.'

And as they moved towards me, I thought I was seeing things. There were people in wheelchairs, people on crutches, people moving with the aid of walkers and canes, people hardly moving at all. There were people old enough to have been my grandparents – and they moved with the kind of hurry you usually find in tortoises. From the way these people had been sitting and standing, I'd had no idea. I hadn't seen a walker or a wheelchair in the crowded room, and I hadn't seen anyone hunched over or struggling, and now I was scrambling to make sense of this unexpected turn. I caught the eye of one elderly man, wheeling his own wheelchair, coming right towards me, and I thought, 'Oh my God, what are we into here?' I looked into as many faces as I could, and what I got back were expressions of help-lessness, and weariness, and gratitude. All at once, and all mixed together. They were happy to see us, and desperate to get out of there.

What had happened, best we could puzzle together afterwards, was that these slow-moving office workers had been directed to this one office some time earlier, after they had about run out of gas on the way down. It's possible their rerouting coincided with the realization that the lower stairwell had filled with debris when the south tower came down. Whatever it was, some of these folks had climbed down as many as 70 flights of stairs to this point, and they were beat. Thrashed. They'd also been slowing up the works on these lower floors, and had perhaps been ushered aside to make way for the faster-moving people. Or maybe they were simply catching a rest before continuing. With them were their friends or co-workers or kind souls they'd met on the climb down who'd been helping them. The folks in wheel-chairs had been getting lifts, one landing at a time. This, in itself, was remarkable, that people would stop to help strangers in need, in just this way. The folks on crutches each had a couple of people with them to assist. Pretty much every non-ambulatory civilian had at least two able-bodied friends in tow. It was an uplifting thing, once the picture came clear to us, the way so many people had put their own evacuations on hold long enough to help these other people negotiate the long, difficult climb. A few people actually introduced themselves on their way out the door; they called out their names and the names of their friends who were helping them and

Win a Lotto
IT COULD BE YOURS!

This card could be worth £50,000!

Don't throw it away!
3 MATCHING SYMBOLS GUARANTEES
YOU ONE OF THE REWARDS OVERLEAF

£50,000
HAVE YOU WON?

2 matching symbols will entitle you to a gift

Game 1

Game 2

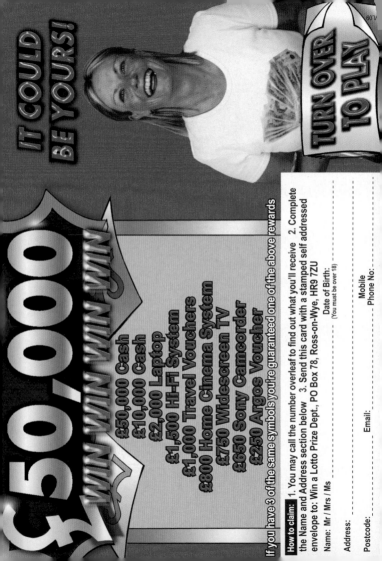

IT COULD BE YOURS!

TURN OVER TO PLAY

V109

£50,000

WIN WIN WIN WIN WIN

£50,000 Cash
£10,000 Cash
£2,000 Laptop
£1,500 Hi-Fi System
£1,000 Travel Vouchers
£800 Home Cinema System
£750 Widescreen TV
£650 Sony Camcorder
£250 Argos Voucher

If you have 3 of the same symbols you're guaranteed one of the above rewards

How to claim: 1. You may call the number overleaf to find out what you'll receive 2. Complete the Name and Address section below 3. Send this card with a stamped self addressed envelope to: Win a Lotto Prize Dept., PO Box 78, Ross-on-Wye, HR9 7ZU

Name: Mr / Mrs / Ms _____

Date of Birth: _____
(You must be over 18)

Address: _____

Postcode: _____   Email: _____   Mobile Phone No: _____

what floor they were on when they started out. And I stood by the door, collecting these identifications like they were about the last traces of goodness we might find in the wake of such destruction. It was the one time that morning I nearly cried. The tears would come later, but these various shows of extreme human kindness just about sent me reeling. I set the heroic, selfless efforts of these gracious people against the evil terror that had brought about these attacks and I allowed myself a small smile, which seemed to me a more fitting response than tears. Like I said, it was an inspiring thing. Uplifting, even, and I smiled to myself again as I realized the word applied in every sense, as some of our stronger guys hoisted the several wheelchairs, one man on each wheel, and moved quickly down the hall to the stairwell.

I didn't do a headcount, but there had to have been a couple of dozen non-ambulatory office workers in that room, easy, and each one of them had one or two or sometimes three people helping them. My first move, then, was to get rid of the helpers. I didn't mean to appear crass about it, because these good people deserved our respect and appreciation, but we had the assistance part covered. There were about 60 firefighters lining this one hallway. They were trained for this kind of thing. I thought we should get these able-bodied people down so that we could worry about everyone else, but these people were reluctant to leave their friends. I can understand that, now, and I suppose I could even understand it at the time, but I refused to accept it. I had a few folks telling me they'd helped so-and-so down from the 50th floor, or someone else from the 60th floor, and they weren't about to abandon them after all that, and I had no choice but to be hard about it, cold. 'Great,' I'd say, 'you've done good, but now it's time to go. We'll take over from here. You need to get down.' I didn't have the time to take time with these good people, or the patience to be patient. I had a job to do, to clear that floor as quickly as possible, and there was no room in the assignment for me being a nice guy.

So off these kind souls went, no doubt mumbling to themselves about the asshole chief in charge of the evacuation, and we were left with about 20 people in a variety of situations. Wheelchairs, walkers, crutches, walking-casts... the works. The wheelchairs were easy. We could roll them down the hallway, down the stairs even. They moved along pretty easily. It was a bumpy ride, going down those stairs, but it went fast; I'm sure it wasn't comfortable, but there was no time for comfortable. It was the folks on foot who gave us some trouble. We had a couple of stubborn old bond-trader types, who said things like, 'I've walked all this way on my own, all the way from the 58th floor, and I mean to walk the whole way myself.'

Once again, I had to be cold. 'No, you're not,' I'd say. 'We're gonna give you a ride.'

Whenever possible, we looked to the interior offices for chairs we could use to carry some of the heavier individuals. I'd put four, five, six rescue workers on one non-ambulatory worker, thinking they could take turns carrying them, two to a man to each landing, four to a man if they were ferrying the individual on a chair, then they could switch off. Some of these civilians were overweight, pushing 250, 300 pounds, so I quickly matched them with what I took to be my strongest guys. It was a real seat-of-the-pants kind of dispatch, assigning this one group here, this other group there... whoever caught my eye, I gave them an assignment. And there were more than enough helping hands to go around. Way more. It got to where I still had 20, 30 guys with nothing to do, and at this point in the evacuation they were pretty much itching to help somebody, anybody but themselves. 'Come on, Chief,' I'd get back, 'put us on with one of these people.'

'Go,' I'd answer. 'You want to help, just go. Clear the stairwells.'

I didn't tell these matched-up rescue workers anything they didn't already know, but I told them anyway. I told them, if they were carrying someone in a chair, to wait until the half-flight of stairs beneath them had cleared to the next landing. Made more sense to do their waiting on the landing than on the steps. I told them to switch off, as often as possible. I didn't want anyone getting tired. We had plenty of manpower, but we still had plenty of floors to clear.

In this way, I matched each disabled, elderly or just plain tired office worker to an appropriate group of firemen, and sent these groups on down. The last person to leave that glass-walled office was a 59-year-old Brooklyn grandmother named Josephine Harris, a Port Authority bookkeeper who was completely spent from her tough descent from the 73rd floor. I matched her with an entire company, Ladder Co 6 out of Chinatown, which was captained by a good friend of mine named Jay Jonas. Jay and I used to study together, and firefighters spend about as much time on the books as medical students, committing all kinds of technical and scientific and architectural information to that place in our memory banks where we don't have to think about it to call it to mind.

We'd spent countless hours together – I'd been in his house, he'd been in mine – and we knew each other pretty well. So well, in fact, that I knew I didn't need to say anything to Jay about taking good care of his lady, but I told him just the same. There was something about Josephine that seemed deserving of my extra-special attention, and Jay's, and that of his men. Couldn't quite put my finger on it at the time, but she seemed special, the proud way she had managed her tough circumstance. Far as we could tell, she'd be the last civilian to make it down from that building, and that was something worth celebrating, worth taking good care of over these final floors.

Let me tell you, the climb down hadn't been easy on poor Josephine Harris. Her legs had swelled. She was having some trouble breathing. Hell, she could barely move, and yet when we first started out from the 12th floor she wasn't really slowing us down too much, because there was a kind of accordion effect happening in the B stairwell. There had been so many of us moving down that narrow stairwell, with such an intense initial push, that to start with Josephine's few steps at a time were enough to keep pace. But that soon changed, as the stairways opened up and the people were hurried along. These firemen had done such an incredible job getting that roomful of people out and down, that there was now a full flight of stairs ahead of us, all clear, and Josephine's deliberate pace became more of a concern. Or, I should say, it became more noticeable. We weren't really concerned at this point. We were very nearly there, and we knew we'd get there eventually.

Trouble was, eventually was taking its time. Josephine was moving, and Jay and his men were moving with her, but they weren't making any real progress. Me, I'd pull up behind them after clearing each floor, and stand and fidget until I could snake my way to the next re-entry door. I got with Jay as we made the return between the 11th and 10th floors, and we figured the thing to do was grab a chair to assist with the effort. Josephine was a proud woman, and it wouldn't have worked to just hoist her in a fireman carry and keep moving; I don't know that it would have worked anyway, so as I did my sweep of these next floors I started checking, real quick, for a suitable chair. I'd been looking all along, but now I looked harder. As we dipped into the single-digit levels, though, the layouts of each floor got a little funny on me, and I quickly realized why. The north tower had that massive sky lobby, reaching up from the concourse level, which left some of the lower floors with a lot less room for office space. This didn't mean a thing for all these people coming down in the stairwells, but it meant a great deal if you were looking to find a chair lying around not too far from the re-entry door. I didn't have time to run to the perimeter offices, where all the big bosses had their nice big-boss office chairs, but all I could see in my limited view were those flimsy secretarial chairs, the ones with no arms and hardly any back to them, and Josephine was better off on foot than on one of those babies.

Jay and his men were continuing down with Josephine, and I was doing my hurry-up-and-wait routine, sweeping each floor, catching up to them on each new landing, and each time I caught back up Jay would flash me a look that seemed to say, 'Where the hell's the chair?' He was looking also and we both were coming up empty. Believe me, I was frustrated that I hadn't found a solid, armed, straight-back desk chair, but I had to weigh my search against the time I'd lose in the searching.

Anyway, we were moving. We weren't making any kind of time, but we were moving. Jay and his men were half-carrying Josephine, half-supporting her. She'd take each step with both feet – meaning, she'd step down with her right foot, place her left foot alongside, and then rest for a beat before attempting the step just below. Every now and then, she'd ask whoever it was at her side if she could catch her breath for a moment, and whoever it was wouldn't have the heart to tell her it wasn't such a good idea. And, so, she moved, in stops and starts, from ten to nine to eight. Her pace became Jay's, and the rest of the guys from 6 Truck; their pace became mine. Soon enough, with the kind of crawl we were doing, a huge space opened up between us and the rest of the pack from that 12th floor office.

It was about 10:29, some seven or eight minutes after we'd come upon that hard-to-figure collection of variously abled office workers, and I imagined they were all down by this point. I thanked God for that, but at the same time I was wishing us on to that lobby floor with them. For some reason, the word 'painstaking' popped into my head, and that's really the best way I can think to describe our progress. And it was hardly progress. None of us begrudged poor Josephine Harris the painstaking effort – in fact, she seemed blessed to have made it this far, and I'm sure these guys all felt blessed to have a hand in her making it the rest of the way – but I'm also sure that all of us wanted desperately to be out, and down, and done. To breathe fresh air. To get home. Josephine included.

We'd had enough, no question.

But there was more to come. Much more. As I raced through the stairwell, between the seventh floor and the sixth, I heard that noise again. That same sick, killing rumble from just 29 minutes earlier. There was no mistaking the roar, and as it quickly approached I knew what it meant. We all knew what it meant.

SIX: Collapse

I hadn't counted on the wind.

It was loud the first time, but this time it was ear-splitting, bone-chilling, knee-trembling... every-damn-body-part-shaking, all multiplied by about a million. And this second blast was topped off by a frightening wind, so I suppose it was hair-flapping as well.

The resulting storm of noise was so far beyond loud there needs to be another word coined to describe it. If there is such a thing as exponential noise, of sound and fury that increases by uncountable multiples – beyond comprehension, beyond registering – then this certainly qualified.

Man, it was howling! No way to accurately describe it, really, except to know it first hand, to be *inside* it, and I wouldn't wish that perspective on anyone. Best I can manage is that it was like being caught in an avalanche during a hellish tornado while trapped in a wind tunnel, all beneath a crashing tsunami, with no hope of escape. There, that about says it. I don't mean to overstate, or exaggerate, but the experience left me feeling no words could do it justice, so I reach for what words I can find. Maybe an image would do: mount a video camera in a bunker, set the whole thing off with a couple of hundred pounds of explosives and if you could retrieve the tape you might get back pictures to reflect what it felt like in the eye of such a killing thunder. More likely, though, you'd have no idea.

The wind seemed just shy of gale force, and I learned later it was caused by a mountain of compressed air tumbling down the tower as it buckled in on itself, and as it came roaring, ripping towards me I knew what was coming. This, I'm thinking, was the worst part of the rumbling stampede, the fact that I knew what it meant. There were guys in that stairwell who still had no idea the south tower had come down, who likely had no time to process the sudden wind and noise for what it was, and for them, I suppose, these next moments were unsettling without also being terrifying. But not for me. It was 29 minutes after ten o'clock in the morning, not a half-hour since the first tower had fallen, and I suppose if I hadn't just lived through the same muted roar across the plaza, and learned of the destruction that had come with it, I would have saved myself a few seconds of complete anguish and fear. As it happened, though, I knew. In a split second, I knew. And I knew for certain. There was no mistaking these sounds for anything else. There was no mistaking the sick rush of wind. There was no mistaking the violent, all-over shaking. It was all terribly, terrifyingly clear. I even found time to give these thoughts voice, I learned afterwards, because as it registered what was happening, I blurted out, 'Oh shit, here it comes!' Not exactly the most memorable parting phrase in the history of last words, but I said it without thinking. In fact, I don't even remember saying anything, but a couple of guys told me later they could hear my voice where they stood, on the sixth floor, or the fifth – and for the most part they had already come to the same conclusion.

In an instant, the whole of my life washed over me. Just like they say in the movies, in cliché. My wife and my kids. Boom! There they were. Instantly, there was this life-affirming picture of them, just burned into my brain. Didn't take more than a beat to present itself. It wasn't a shot from our refrigerator or a family photo album, wasn't a pose any of them had taken or that I remembered seeing, but there they were, flash-frozen, front and centre. Smiling. Thank God they were smiling, I thought, and in the same instant that I took in this picture, I contented myself knowing that at least they'd be taken care of. Here on in, they'd be okay. There was no time to think things through, but time enough for this. Right then, from that first split second, I thought what their lives would be like with me gone. I was sad, and grateful. Sad that I wouldn't be there to see how things turned out for them, sad that Stephen and Lisa would go the rest of their way to adulthood without a father, sad that they would grow to be parents without me cheering them on, sad that Debbie would be alone – but grateful for the time we did have together, grateful that they wouldn't have to worry. Financially, they wouldn't have to worry. The department would take care of them. My full salary, tax free, every year for the rest of Debbie's life. Full benefits. All kinds of other one-time payments. Insurance benefits. Tuition assistance for the kids. They'd be okay. I'd be gone, but they'd be okay, and I ran the numbers and their futures in my head in the time it took for me to think, 'Oh shit!' I processed the whole damn scene – theirs, mine, the firefighters' down below, that poor woman that Jay and his guys were helping down this same stairwell. I could see it all, and it wasn't good.

Amazingly, though, I found I wasn't scared of dying. This surprised me, a little bit. I'd always said as much, whenever I'd been asked, and when you're a firefighter you get asked this question a lot, but I'd never really had to put the question to the test. Hadn't been made to think about it in such a concrete way before, but here it was, full in the face, and it didn't seem so terrible. Just another part of the drill, and what came to me in fact was what I'd expected in theory. No, I wasn't afraid to die. What scared the shit out of me, though, was suffering. That, to me, was worse than death. I couldn't even imagine it, the pain that was coming my way, the enormity of it. I thought to myself, 'Whatever it is, just make it quick. Please, God, make it quick.'

And then I prayed. Formally, informally… every which way I could think to pray, I prayed. See, the praying came easy, natural. I was born and raised a Catholic. I had spent eight years at Catholic grammar school, and four years at Catholic high school. I took theology courses in college. I am, for better or worse, a God-fearing man, a praying man, although I wouldn't cop to being the most devout Catholic prior to that September morning. I went to church grudgingly, infrequently. To set a good example to my children. For appearances. To do the right thing. I guess you could say I was a practising Catholic who had fallen out of practice, but as that black rumble came hurtling down towards me I was born again. So I started praying. Hard. Fast as I could. Making up for lost time. Making full use of the time I had left. And yet for the life of me I can't remember now what prayers I was saying. Either the Hail Mary or the Our Father. One of those. Who knows, maybe I managed snippets of both, but the words came bubbling out as if by themselves. I'd said these prayers so many times they were a part of me. I didn't even have to think about it. It just came…

Our Father, who art in Heaven…
Oh shit, here it comes!
Debbie.
Please, God, make it quick.
Holy shit!
Stephen, Lisa…
Hail Mary, full of grace…
Please, God, make it quick.
Oh shit, here it comes!
Debbie, and the kids…
Our Father, who art in Heaven…

My thoughts were all over the place, one on top of the other, all at once, over and over and over. There was no time to reel them in or to make sense of them. They were just there, competing for my attention. Jockeying for space in my head. Or maybe it was some kind of defence mechanism, all these thoughts distracting me from what was happening.

Somehow, underneath all this frantic thinking, I landed on a course of action, which is interesting to me now, after the fact. Some people, I'm sure, would have simply curled up in a tight ball, the way we used to have to do for those air-raid drills back when I was a kid in school, during the Cold War. Remember that pose? It's the position of defencelessness. Hole up and die. I think about this now, and it seems to me this is probably how a great many folks in those towers met their fates. They covered their heads and waited for whatever it was to find them. Others might have stood stock-still, unable to move. Others might have

The white circle indicates B stairwell,
in which Richard Picciotto was trapped.

Richard 'Pitch' Picciotto

From left to right:
Brian Kenny, Artie Leecock,
Capt. Jack DaSilva, 'Proby',
Richard Picciotto, Dan Paisner,
Tim McCarthy and Dave Koyles.

THE OTHER TWIN TOWERS OF NEW YORK

NEW YORK POST

The *Post* has been inundated with phone calls and e-mails.   Starting this Monday, Sept. 24, The *Post*'s Web site,

screamed, or cried, or panicked in some way. There's nothing wrong with any of these responses, or any other response for that matter. They're just reflexes, really, that's what I've come to realize. Knee-jerk reflexes to impossible situations. At a time like this, when the end is pretty much certain, we do what we have to do. We do what comes.

Me? I ran. I bolted down those stairs, between the seventh floor and the sixth floor, like my life depended on it – which, in a very basic way, I guess it did. It wasn't a logical move, but it was a move. It was some way to take action, something to do. I wasn't the kind of guy to go out without a fight, and this was my fight. It was like that fight-or-flight reflex I wrote about earlier, only here it was fight *and* flight. It was every instinct rolled into one. I wasn't running away from the obliterating force that would find me in this stairwell so much as I was trying to *out*-run it. Does that make any sense? It wasn't a cowardly act, I don't think, but a desperate one – one that would have called on superhuman feats of strength and speed for success, and the superpower to stop time, for good measure. But there was no fleeing from this cascading terror, no logical moves to be made. The building was coming down in a torrent, and I would be forever stuck in its middle – no matter how fast I ran, no matter where I turned, or what prayers I mumbled to myself.

Still, I ran. Hard as I could. Two or three steps at a time. Recklessly. All told, I couldn't have moved more than a dozen steps, there wasn't time, but I moved. All the time thinking, 'Please, God, make it quick. Please, God, make it quick.' Like a mantra. 'PleaseGodmakeitquick. PleaseGodmakeitquick.' Over and over and over, and as the building shook violently and the roar become louder still, I realized I could hear myself breathing. Through all of this, I could hear myself breathing – amazing! – and I even found time to worry which breath would be my last.

I was on the landing between the seventh and sixth floor when it hit. At least, that's where I place myself in recollection. I could have made it all the way down to the sixth, but I seem to recall pulling up on the return landing, about to zag back down the next half-flight. All along the building had been shaking, rocking like crazy, and here it was shaking so hard it was an effort to keep my balance. Beams started falling. Big chunks of concrete. All kinds of shit, just raining down on me, like a friggin' plague. Raining down on *us*, I should say, because I wasn't the only one on those stairs. Even now, at the very end, I didn't stop running, hurtling down those stairs with everything I had left, making the turn on the landing. I pressed on, until finally a beam or a plank or some blunt object hit me on the head and sent me reeling. Knocked my helmet off, too. Whatever it was had whacked me pretty good, and I was down and thinking that would be it.

But that wasn't it. The noise kept coming. The debris kept falling. The building kept shaking. And I was still breathing, and conscious, and alert to the whole damn thing.

I tried to stand, but stumbled before I could right myself, and in the stumbling I cursed the fact that I was still alive. Really, I was pissed. And afraid. I should have been dead by now. I'd hoped I'd be dead by now. So much for praying it would be quick. I'd thought it would be instantaneous, however it was gonna happen, but this was taking way too much time. This was torture.

All of this unfolded in a blip. The south tower had taken ten seconds to come down, and this one would take only eight seconds, according to published reports after the attack. Once again, eight seconds doesn't seem to cut it, in my estimation. It had to have been longer. Three or four times longer. How else to explain these runaway thoughts, these stops and starts as I made to escape, these twists and turns? But I suppose there's no explaining it, this stretching out of time.

I did manage to stand, finally, only as soon as I got to my feet the landing fell away. Like a trap door. It was there, and then it wasn't there, and as it fell I moved with it. The weird thing is, I don't really recall hearing any noise at this point. Either it was so deeply ingrained and so much a part of this harrowing scene it had quickly become like background music, to where I no longer noticed it, or it had become quieter. This last makes sense, when I think about it, because by this point much of the wreckage had settled into the sub-basement and surrounding area; by the time it reached my little landing between seven and six there was hardly any noise left to be made. And you know how it is at the silent end of a great racket. There's this eerie hollow, this audible emptiness. This giant nothing-at-all where there had only recently been everything, and too, too much, and it was into this space that I fell as well.

Please, God, make it quick.

Actually, it was more ride than fall, more slide than drop, because as the landing gave way, the rubble it became kinda cascaded and skated along, bumping up against walls and stairs and railings and beams and whatever the hell else was in its path. I was a big piece of debris, hurtling through the shell of that stairwell. There was a brief stomach-dropping falling sensation, mixed with all kinds of banging and tumbling, and then a descent in slow motion, and the reason it felt like slow motion, I later figured, was because it wasn't a sheer drop. It *was* slow motion – sort of. Anyway, it was slower than I would have thought. It was like being on the rough end of a relentless rockslide. Someone suggested to me later it was probably like getting caught in unbelievably rough surf, like when you're body surfing and a wave sneaks up and knocks you silly and over and forward, to where you no longer know which way is up or how to arrest your fall. But that wasn't it. Close, but not quite. It was more like a short free fall, where you never fully leave the ground, because the ground is free-falling with you. Tumbling, as if down a flight of stairs, but as if I was taking the stairs with me. Hard to explain, and equally hard to understand when it's happening to you.

At some point during this wild ride, the lights went out. I didn't notice them going, and it's possible I had my eyes closed to what was happening, but when I opened them there was nothing to see. Just blackness, all around. Nothing more. The kind of blackness you need to raise your arms against, to cover yourself, for protection, because you've got no idea what's coming, or from what direction. But there was nothing coming, nothing still falling. There was no more noise. It's like those stupid rides they've got in shopping malls, where you put a quarter in the slot and sit your kid on top and the thing shakes and shakes for a half-minute or so and then the quarter runs out.

And here's another weird piece: I didn't land. There was no big trauma, or impact, when the falling stopped. It just stopped. I had my bunker gear on, which offered some degree of padding and protection, but no way was I impact-proof; I was getting whacked and paddled and smacked around pretty good, and then all of a sudden the whacking and paddling and smacking sort of stopped. It slowed, then it stopped. It lasted for a good few seconds, and at the other end there was nothing. I didn't really have a single clear thought, as I tumbled. I was just waiting on the end. Disorientated, and waiting on the end. My end. It hadn't claimed me before, that first whack that had knocked me to the ground, but surely I couldn't survive such as this. Here again, as before, my head was filled with every possible thought, every conceivable outcome, every unfolding detail, and somewhere in there too were the stuffings of my life, the small moments I couldn't shake, the *what if*? scenarios we all carry. As I fell, I expected my thoughts to slam shut on me with some final impact. There'd be thinking, and feeling, and then there'd be no thinking, no feeling.

Please, God, make it quick.

But nothing happened. I stopped moving, that's all. I stopped thinking in any forward-thinking kind of way. I stopped *being*, if that makes any sense, for a beat or two anyway. I didn't black out or anything, never lost consciousness, but I did momentarily lose a full grasp on where I was and what had happened and what I was doing. There had to have been a couple of ticks of the clock in there where I was merely still, and not thinking, and unaware of any single thing. Understand, I couldn't see. I couldn't grasp my situation. I couldn't move, once again for fear of making the wrong move. Plus, as I began to assess the scene, I didn't think I had any ability to move, or any right. Or even anything to move. I thought I was dead. I actually wondered if this was what it felt like to be dead. Think about it: it was pitch dark. There was no sound, no movement, no nothing. My mind was still working, but to someone who'd been raised Catholic, to someone who believed in the concept of Heaven and Hell, this didn't strike me as too surprising. Out of my experience, perhaps, but not out of the realm of possibility.

Okay, I determined, so I could think. For a second, I even allowed myself a secret sigh of relief that the memories I'd made up until just that moment wouldn't die with me. I could still think and feel... and *remember*. This wasn't so bad, I thought, this being dead. I'd thought I'd be above the clouds, floating and flitting about with angels, looking down from above, but maybe this is what my afterlife would look like. In the beginning, anyway, this was how it would be. The clouds and the angels and the somersaults would come later. So, no, this wasn't too bad. Wasn't how I'd pictured it, but it wasn't too bad. And I hadn't really suffered, beyond the unbearable fear of suffering. It was over, and yet it wasn't over.

Gradually, though, I began to notice my body, and as I did I took time to realize that if I had been dead I wouldn't be noticing my body. There'd be no body, right? There'd be no arms to move, no head to turn. But here it was, my body, very much intact. I could *feel*. I could squeeze my fingers into a fist. I was lying on my back, not quite flat, splayed across an uneven surface of rubble and cement and uncertain debris. My head was resting on a cement boulder, which appeared to be about the size of a softball. There were no beams crushing me, no piles of concrete. I had no idea what sort of space I'd been dropped into, but nothing was confining, or constricting me, not that I could tell anyway. I could feel with my hands that I was lying among a mess of building materials that had been reduced to pebbles, stones. And, mostly, a very fine, fine powder. I reached around, to get a feel for where I was, and my hands noticed that I was pretty much covered with the stuff. Ash, dust, powder... whatever it was. It was in my mouth, in my nose, in my eyes, in my ears. Everywhere there was an opening, there it was. It was beneath my clothes, even. It was all around. I was buried beneath a coating of about six inches of the stuff. Buried alive. It's a stupid phrase to use in this particular instance, because even a coating of six inches of the stuff can be shaken off without too much trouble, but it's what popped into my head at the time. No, I wasn't buried to the point where I couldn't claw my way out, but I was covered. Completely covered. And I was alive!

This was the most remarkable part. I couldn't believe it. Yes, absolutely, I was still alive. When my mind caught up with what my body was feeling, though, it didn't really get it, not at first. I couldn't catch up with my thoughts, with the slow realization that the collapse hadn't killed me. It all came back to me – the rumble from on high, the recollection of what had happened across the plaza, the smoking infernos against the clear blue sky, the commotion of the makeshift command post in the lobby, the

headlong race down the stairs to try to beat the fall...
I thought, 'Something's wrong. Something's got to be
hurting.' I understood what had happened. I understood what was happening, still. And I understood
that I was still breathing. What I couldn't understand, though, was why I wasn't in any kind of pain.
I had to have broken something, or be bleeding
internally, or have suffered a concussion, some sort of
open wound... something. But I felt fine. Or, mostly
fine. I was choking, coughing. My eyes were stinging
like crazy. I had a couple of scrapes and bruises, here
and there, but basically I felt fine. Nothing more than
sore.

I'd lost my helmet in that first fall, so there was all
this dust in my hair. Mounds of it. Imagine dumping
100 pounds of baby powder on your head, and that's
what it felt like to me, and as I propped myself up on
my elbows and shook out my hair I got another face-
full of the stuff on the rebound. Man, trying to shake
that powdery ash out of my hair was something! I
could feel it, realighting on the back of my neck, on
my cheeks, everywhere. I couldn't shake it loose. All
I was doing, really, was moving it around.

I sat all the way up, and did a quick inventory on
my body. I touched myself, head to toe. Everything
appeared to be in place, in good working order. I
didn't appear to be bleeding, in any kind of open-
wound sort of way. I couldn't see a blessed thing,
though, so I was going on touch. Touch and sense. I
couldn't hear anything, either, but all of a sudden I
had the strangest feeling that I wasn't alone. It kinda
crept up on me, like a slow realization. I went from
feeling completely alone to feeling comforted by
some unseen company in such a way that I didn't

notice the shift. I didn't think to speak, not just yet, but it was the oddest thing, the way I sensed there was someone else with me, wherever I was. One someone, two someones... I had no idea how many, but I wasn't alone.

I was alive, though, this much seemed clear. And not too happy about it. I'd gotten from there to here on borrowed time, *there* being the moment I caught the first tremble and knew the building was about to come down and *here* being right here, right now, right at this very moment when I was trying to piece together my circumstance in a way that made sense. It was all borrowed time, and it was more than I'd bargained for. More than I needed. Hell, more than I wanted. Way more than I wanted, because just as I assessed my own situation I also assessed the situation in general, and I realized that if the entire building had indeed fallen, and if I had indeed survived the collapse, then this had to mean there were over 100 floors of concrete and debris and mountainous rubble on top of me. Thousands of tons of twisted steel and shattered glass and bent-out-of-shape office furniture and no way in hell to break free.

Buried alive.

There was that phrase again, only this time it meant what it meant in the headlines. Buried. Alive. Entombed beneath a man-made (and man-destroyed) mountain of steel and concrete. Never again to see the light of day, or to breathe fresh air, or see my kids or hold my wife. It was, I feared, a fate worse than death, because it meant I would be here a good long while. It meant I would suffer. I hadn't suffered yet, but it would come. Slowly. I'd starve, or go quietly crazy, or suffocate. Or maybe there'd be some kind of fire and I'd burn to a crisp. Or a secondary collapse, sent to crush me like a pancake. I thought about where I was, and what the devastation must have looked like from outside, and what it meant to be trapped beneath all that devastation, and what I'd survived to make it to this godforsaken point, and I could think of no good outcome. Not one.

I knew I'd somehow fallen into some sort of void. I'd made my living studying building collapses, so I knew it wasn't at all unusual in a high-rise collapse for people miraculously to survive, simply by being in precisely the right place at precisely the right time. (The phrase didn't exactly apply in this case, because the right place at this particular time in history would have been several miles away, but you get the idea.) It was impossible to predict where or how these voids would form, but there was always a pocket or two, somewhere in the rubble pile, and I'd stumbled into one without meaning to.

Please, God, make it quick.

But here I was, looking at a shitload of suffering, and I thought, 'So much for praying. So much for checking out on my own terms.' I've got to tell you, there was a real black moment in there, during which I went back to feeling so completely and thoroughly alone, like there'd be no one to record these last moments I was now facing, however long it took for these last moments to play out. For some reason, this struck me as profoundly important. That I would survive the collapse of the tallest building in the world, from the inside, and that I would likely survive another couple hours or even another couple days, and that *no one would ever know*! My wife and kids, my parents, the rest of my big family... My brother firefighters. No one would know what I was knowing now. I thought this, and then I wondered, 'What the hell was the point?'

And then I got that feeling again, that sense that there were other someones in there with me. All of this happened in a matter of seconds – fractions of a second, even – the bouncing back and forth between thoughts, the collecting and tossing aside of different ideas, the trying to determine if I was dead or alive, the feeling alone and feeling connected. It came and went, like a wave, but now it was upon me again, and I felt certain there was someone here with me, facing the same thoughts of gloom and doom, caught inside their own silences. There were no sounds, no rustling, no sense of movement. But there it was. It had come upon me as soon as I'd come to, and then it left me for a while as I figured things out, and here it was again.

The best way to answer a fear or a hunch is head-on, right? So I called out, to anyone in earshot. 'Is anybody here?' I hollered, into the blackness. It was a long beat or two, then I put it out there again: 'Is anybody else here?'

For the life of me, I couldn't tell you what I was expecting to hear in response, or what I was hoping to hear. Something, though. Anything. Anything but the sound of my own voice.

SEVEN: Void

'Is there anyone here?' I said into the blackness. 'Can anybody hear me?'

My voice was choked with ash, and I suppose also with emotion, and in the deepest part maybe a little fear, and when I heard it back it sounded as if it had been swallowed up by what was left of that stairwell, by the cushion of dust. Like I was speaking into a pillow. It didn't sound like me at all. I had no idea what I was hoping to get back, but I was praying on something. Even the certainty that there was no one else around would have been a notch or two better than the eerie sense that I had company.

Finally, after the longest couple of seconds in the history of long couples of seconds, I got a reply. 'Yeah, yeah, I'm here,' I heard. The voice seemed to come from down below. It also sounded swallowed up, by this peculiar pit into which we had fallen. It wasn't faint, the voice, but at the same time it wasn't a thundering holler. Tough to place, how far away it was coming from. But it was a goddamn welcome sound, I'll tell you that much.

Then, 'I can hear you.' It was another voice, also from down below, also tough to place. Both voices seemed to come from about the same distance, from about the same direction.

Finally, from up above. 'I'm here, too. I can hear all of you.'

Three different voices, added to my own. Three different versions of the same bad scene. A regular friggin' chorus, and it didn't end there. There were more voices, checking in – and reports of a couple more survivors, out of my earshot but in an adjacent void. I'm guessing it took each of us about a minute, perhaps a little less, to run through our own personal triage, to assess our circumstances, to determine for ourselves that we had somehow managed to survive the collapse, and that our prospects were as black as the hole into which we had fallen. Here again, in a morning filled with segments of time that seemed to race ahead, and other segments that seemed to stand still, this stretch felt a lot longer than it actually was. For the longest time, I'd worked to get a clear fix on what had happened, on what was happening still – and yet this longest time unfolded in no time at all. Sixty seconds? Man, that was nothing, and into it I'd placed every possible thought, every conceivable scenario, every workable outcome. Every damn thing, and then a few more besides.

I couldn't see a thing. Black as coal. Couldn't make out any movement, any flicker of shadow. Nothing. I'd lost my flashlight in the rolling tumble, and I patted the jagged surface at my sides, hoping it would turn up within reach. Wishful thinking, I suppose. These others, I imagined, were doing something like the same, because there were no beams of light anywhere in my view. At a time like that, if you had a light, you would have flashed it on, no question. First things first, right? You look around if you can, soon as you can, catch your bearings, start hatching a plan. As it happened, I did have a light, a tiny 'mini' Mag flashlight I always kept on my belt, but I didn't think to reach for it just yet; I guess, in the calm after the storm, I wasn't thinking clearly enough to think of everything. The hand-held light, that's what I was wishing on.

Slowly, as my eyes adjusted to the blackness, my attention was pulled to a small bar of light below me on the stairway. Actually, to call it a light was to give it more power than it really had, because it was little more than a hardly bright image in the middle of a black landscape, and it certainly didn't do any lighting. It gave off a glow, but I still couldn't see a damn thing. It was more ember than light, and it served more as a fixed point than a beacon, but I stared at this small beam of light as if it held some great power over our fate. I figured it right away for what it was – an emergency fluorescent light from the stairwell, still hanging on and doing what it could – but it was so far away and covered with so much dust and muck that no real light could penetrate. It hung there like a sliver of a star, in deepest space, and I was no better off for having discovered it, except for the momentary ray of hope it offered.

Clearly, we weren't moving or getting out of there on the back of that light. I knew this, even as it held my gaze for a good long time. The thing to do, I thought, turning away finally from the dim beam, was a kind of roll-call. Forget the light. Forget the seeing clearly. We needed to figure out who we had, what kind of shape we were in, what kind of tools we still had in our possession or vicinity. I was guessing we were all firefighters, from the fact that we were in the stairwell at this late point in the evacuation. I'd eye-balled everybody, pretty much, on the way down, and the civilians were long-gone when the death rattle hit. At least, that's what I was hoping for. 'This is Rich Picciotto,' I said, introducing myself. 'Battalion Chief, Fire Department, 11th Battalion. Who else is here?'

One by one, they sounded off – starting in my immediate vicinity. Mickey Kross, a lieutenant from Engine Co 16. Jim McGlynn, a lieutenant from Engine Co 39, who had three of his men with him, a 'proby' named Rob Bacon and two senior firefighters, Jeff Coniglio and Jim Efthimiaddes, trapped in a separate but adjacent void, just below the landing one half-flight down from where I sat. Our voices worked their way up the stairwell, past my position on the half-landing between the third and second floors, all the way up to my friend Jay Jonas, captain of Ladder Co 6, who'd come to a stop two to three storeys above me. He hadn't heard me when I first called out my own name to start the roll-call, but as soon as he announced himself my mood picked up a bit. You're not gonna get excited about much of anything in a tough spot like that, but it was good to hear a familiar voice.

'Jay,' I shot back. 'That you? It's me, Richie.'

'Richie?' he said.

'It's me, bro,' I said. 'You okay?'

'I'm okay,' he said. 'I think I'm okay.'

I was too shaken to marvel at the fact that I'd landed in this unlikely, impossible situation with a career-long pal, too focused on the various survival instincts that had by now kicked in to get all touchy-feely-gushy that we would now face our fates together, but it was at least a small comfort to find Jay Jonas in the rubble. I was alone, lost in my thoughts and fears, but at the same time I wasn't *completely* alone, if that makes any sense. There was someone who knew me, someone who knew my wife and my kids, someone who'd been to my house. I wasn't just 'Chief' to Jay Jonas, and this struck me as a great, good thing.

Turned out Jay had his entire company with him – Bill Butler, Tom Falco, Mike Meldrum, Sal D'Agastino and Matt Komorowski – and as they gave their names I wondered how it was that these men had been a couple of floors below me when the building started to fall, and yet here I was, hurtled a couple of floors past them in the collapse. Hurtled past all of them, that is, save for Matt Komorowski, who had actually fallen to a point lower than where I lay. I wondered, too, on the long odds of a whole company surviving such an epic catastrophe. It made no sense, but then this entire morning made no real sense, when I broke it down, minute by minute. Even at this grim moment, with a bunch of us trapped in the carcass of the north tower of the World Trade Center, it was too much to believe.

The thin line that I later learned connected me to several of Jay's guys was also fairly unbelievable. The fire department can sometimes seem like a small family, the way everyone knows someone in common, but in the weeks following the attacks I discovered all kinds of unusual links with these guys from 6 Truck. As it happened, this guy Billy Butler's parents lived in Hawley-Honesdale, Pennsylvania, the same area where my mother was born and raised, the same place I spent every summer, visiting cousins; we knew some of the same people. Mike Meldrum grew up on Staten Island, same as me, and attended the same high school, Monsignor Farrell. Sal D'Agastino also grew up on Staten Island, on the same block where my parents have lived for the past 40 years. That Staten Island connection also ran to Tom Falco and Matt Komorowski.

With Jay Jonas and his men, still, was that Brooklyn grandmother, Josephine Harris, and she appeared to be okay. Stunned, but okay. She didn't add her voice to the mix, but one of Jay's guys sounded off on her behalf, and she hadn't suffered any serious injuries. Below me were those two lieutenants, Mickey Kross from Engine Co 16 in midtown Manhattan, and Jim McGlynn from Engine Co 39, also in midtown. There was also that proby out of 39, Rob Bacon, and a Port Authority police officer named David Lim. I learned later I had a history with McGlynn and Bacon as well; a neighbour of mine went to school with McGlynn, and my daughter Lisa worked as a lifeguard up in the town of Walkill NY with Bacon's brother. In fact, when Bacon first signed on he was hoping to get assigned to a house in Manhattan, and he reached out through Lisa to see if I could pull a couple strings.

Cross, McGlynn and Bacon were all together on a landing below me, and Lim was about a flight above me. Below the lot of us, in a separate void, were those two other firefighters from 39, Jeff Coniglio and Jim Efthimiaddes. Some time later, over the radio, we would also discover another battalion chief, Richard Prunty, from the Second Battalion. We could hear him calling for help. He was obviously hurt. His transmissions weakened and ultimately stopped after an hour or so.

'Don't nobody move,' I said, when the roll-call concluded. 'This place is a house of cards.'

I've since worked to understand how it was that we all fell into command-mode, so soon after the building had given way, and how it was that it fell to me to start off with the commanding. You might think things like protocol and authority would get tossed after a disaster of such magnitude, that it would be every man for himself, pure instinct and adrenalin, but none of us were about to play it that way. I wasn't pulling rank in voicing orders or calling roll, but it was part of my personality, to take charge of a situation. It came naturally. Anyway, I'd never been the type to sit idly by while someone else called the shots, and I wasn't about to start now. More to the point, though, most of us had been trained to follow command. We were soldiers, at bottom, and like it or not I was the highest ranking firefighter in that stairwell. There was Chief Prunty, trapped below, but he was banged up something awful and in no position to take charge – and I was senior to him, anyway. And so it fell to me, this leadership position. I might have grabbed it, and hugged it close, all on my own, but as it happened it fell to me. This was a rescue operation. We were New York City firefighters. We would play it by the book.

Back to the house of cards. That's how I saw the mess we were in. I'd already carried a good mental picture of what had happened across the plaza, to the south tower, about a half-hour earlier, and I'd already allowed my racing imagination to take me to the collapse of this one. I had a clear idea of what had happened, what it must look like, what we were facing – and in truth this was more than mere idea. I'd made a study of high-rise collapses, so I knew we were here on whim and good fortune, and that each could screw us at any time, here on in. Teetering on the edge of a secondary collapse that could be triggered by nothing much more than a sneeze. A giving way of a single beam. We'd be crushed like bugs and no one would ever know we'd survived the preliminary collapse. There was that same thought again, dressed up to include the other souls trapped here with me, and I can't say where it came from, or why it seemed to matter, but there was no shaking it. It seemed a monumental thing, to have made it to this point, and I think we all felt a responsibility to tell the world we were still here.

'Stay right where you are,' I said again, making sure, and then I ran through the roll again, establishing our physical conditions. Miraculously, no one was seriously hurt. Among us, Mike Meldrum had suffered what appeared to have been concussion, and Matt Komorowski seemed to have separated his shoulder, but there were no life-threatening injuries that we could determine. Nothing we couldn't work our way through. Soon as we ran through our sick check, we were on to equipment, and most everyone seemed to have lost his light in the fall. So we felt around at our sides, like drunks looking for contact lenses, combing the debris where we lay. Here again, I didn't think to reach for the 'mini' Mag light I kept on my belt. It was there, but it wasn't a prospect. I was fumbling in the dark, same as everyone else, on my back, trying not to move but at the same time making subtle snow angels with my arms, hoping I'd brush up against a flashlight.

After a beat or two of this, I caught a bright beam down below. Compared to the faraway, dust-covered emergency bulb, this was a friggin' Klieg light.

Someone had turned up a flashlight, or maybe he'd had it on his belt the whole time and had finally struggled it loose and flashed it on. It sounds crazy in the retelling, but we firemen had never been outfitted in any kind of state-of-the-art way. Far from it, in fact. We were the busiest damn fire department on the planet, vested with the safe-guarding of more than ten million people and countless buildings and landmarks and public places, and yet there was no room in our budget for proper equipment. We worked with such a bare-bones budget that a lot of the guys carried their lights on utility belts fashioned from seat belts taken from abandoned cars. In this way, we can strap a couple of tools to the thing and keep our hands free in a fire. Some guys wear their belts as a sash, some around their waists, but most guys wear it, or something similar. We make do with what we can find. Put it this way, a married proby with two kids actually qualifies for food stamps in New York City – so clearly none of us had the kind of deep pockets to afford an item like this, and yet we couldn't afford *not* to wear one.

I don't mean to vent here, but it's incredible to me that even during this grim, extreme situation, we were reminded of the endless cost-cutting and corner-cutting in the department. At least, I was reminded, and I silently cursed our administration for leaving us here without proper equipment. Already, we'd determined that our radios weren't working properly. Our initial calls of 'Mayday!' were being met with static and silence. And now the thought of all my colleagues wearing salvaged seat belts struck me as ludicrous. Who knew what else we would need in the bleak moments that lay ahead? Or, what kind of rescue effort could be mounted to haul our asses up and out of this sorry mess?

Meanwhile, my mini light was in a pouch on my belt, under my heavy bunker gear, but I still hadn't reached for it. I always carried it, but right then it seemed like too much trouble to get to it, and I was fearful of making the kind of all-over movements I'd need to fish it out, so I stayed in the dark. I also carried my Leatherman knife, which usually comes in handy, and a key ring. Here's a small firefighter detail worth noting: every firefighter carries a 1620 key, department issue, which opens every firehouse in the city; it will also control every elevator. It's a master key to the city, really. A lot of times, you'll see a fireman with a key ring on his jacket, with only one key attached, and that's usually the 1620 key. It gets us everywhere we need to go, opens all the fire boxes and fire suppression systems, in all five boroughs. That's one of the four ways you can put out a fire. Smother, separate, suppress, cool. Firefighting 101. When you pour water on a fire, for example, you're doing a combination of all four. I actually thought of this, as I lay there in what was left of that stairwell. The kinds of fires we might face awaiting rescue. The means we might have to battle a blaze from the inside. Chances were we'd be here a while – hours, days, maybe longer – and there was still a lot that could happen. I thought of the jet fuel, and I knew we were vulnerable. Oh, man, were we vulnerable! There was the enormous and unending potential for sparks in a building collapse such as this, miles and miles of cables and wires and water pipes, and the vapours from the fuel could combust at any moment. Remember, we still didn't know what had caused this second tower to come down, had only a half-hour earlier embraced as a possible explanation for the first collapse a bomb of some kind, or another plane. Like most everyone else, I'd long thought these buildings were impervious to most conceivable crashes, or bombings, or natural disasters, or even to isolated fires. Hell, I'd been here in 1993, when that thousand-pounds-plus bomb took out seven mostly sub-level floors, but the building hadn't come close to toppling. Here, though, who knew what we were facing? I couldn't dwell on what I didn't know, else I'd go completely crazy, so the moment these thoughts landed in my racing imagination I moved on to the next ones vying for my attention. Escape. Position. Roll-call. Inventory. Contact.

And, foremost... light. Thankfully, the guy who had a flashlight strapped to his makeshift utility belt had quickly fisted it free and shined it around. Everyone who'd been carrying their lights in their hands had had them shaken loose in the collapse, along with their tools and other gear, but this one light flashed a dusty beam that soon enough led us to a couple more that had fallen within reach, and suddenly I had my bearings. The small area where I was confined was thick with dust and smoke. Even with the light, there was no seeing clearly. My clothes, I now saw, were completely covered with the same grey powder I'd inexpertly shaken from my hair. I looked, I imagined, like a grey ghost – and our situation looked worse. The void into which we had fallen seemed to hold the shape of the pre-existing stairwell, to a degree. The space appeared wider up above, and narrower down below, almost like a funnel. There was a lot of debris. Where I lay, there were no real steps remaining, and what few there were just hung in the air like in one of those Escher drawings. They were steps to nowhere. At other points, though, entire return-flights were intact. Whether or not they would support our weight, we could not yet know, but they appeared intact, rooted. Up above, where Jay Jonas and most of his men sat with the Brooklyn grandmother, there was actually a landing that seemed essentially untouched by the collapse. Down below, it was mostly rubble and beams and girders and pieces of collapsed landing and precious little headroom. (Rob Bacon, who was stranded a couple of floors below me, actually came to rest hugging a giant beam!) The conditions in this one amazing void were vastly different for each group of us, and yet at the moment of discovery none of us really knew what the others were facing. We all thought that where we were sitting was how it was for everyone else.

Initially, I'd counted six people below me, eight people above me, and five or six in the small area surrounding me, but with the light it came clear that I'd double-counted some of them. So that's the first gift the lights gave us, an accurate headcount, and from there we inventoried the tools in our reach. We had axes, hooks, 150 feet of life-saving rope and a Halligan tool. There are very few doors, very few locks, that can withstand a Halligan, so it was an essential piece of gear – and here I hoped we'd have a chance to put it to use. We also had a 'rabbit tool', which is a hydraulic spreader, also used for opening doors. We were pretty well stocked, in terms of equipment, but I knew that wouldn't mean shit if there was no chance to use it.

I hollered up to Jay Jonas, who was about 30 feet above me. 'You see any way out up there?' I asked.

He looked around, for the millionth time. 'No,' he said. 'Tough to see, but I don't think so.'

I hollered down below, and got back the same. Then I got on my radio, but I couldn't raise a signal. A lot of the guys had been shouting 'Maydays' into their radios, all along, from the moment they'd come to, but no one had got a response. It's important to understand here that a 'Mayday' is the ultimate distress call in any rescue situation. I'd never had to use it, in 28 years on the job. None of these guys, with our hundred-plus years of accumulated experience, had ever been on the dispatching end of a Mayday call. A couple of us had heard one or two, we might have even responded to one or two, but we'd never sent one out ourselves – and here we were fat in the middle of the mother of all Maydays. And we were getting back nothing. For some weird reason, I thought back to that famous philosopher's riddle, 'If a tree falls in the forest, and there's no one around to hear it, does it make a sound?' I wondered, 'If 15 people survive the biggest building collapse in the history of mankind, and there's no one on the radio to hear their calls of "Mayday", have they really survived?'

I instructed everyone to go easy on their radios, to preserve the batteries, until we had a better fix on our situation. It wouldn't do us a lick of good to be making all these frantic calls, one on top of each other, not just yet. It was a classic case of redundant effort, so I asked some of the guys to hold off for the next while; the rest of us would have it covered.

Next, on one of my sweeps of the area with

my recovered light, I noticed the re-entry door on the second-floor landing. Or, on what used to be the second-floor landing. I could see the number on the wall, beneath a thick coating of ash and dust, and I worked back from there to piece together my fall. I was about a half-floor above, which meant I'd tumbled from the landing between the seventh and sixth floors, all the way down to the landing between the third and second floors, which in turn meant I'd fallen about 40 feet. Ten feet per storey, that seemed about right. The quick calculation confirmed that I'd taken a significant ride, and that it was indeed a marvel I was still in one piece. Lord knows, the building certainly wasn't. All around me, in these sweeping beams of light, all I could see was compacted debris, beams and girders and chunks of concrete, with no way to tell what our immediate area had once looked like. In fact, for a while in there, I was going on faith that we were still in a stairwell. There were all these stairs around, to be sure, full flights even, but these could have shifted anywhere in the collapse. They could have been lifted, whole, and dropped down at any point; their presence here indicated nothing. There was no logical reason to assume the stairwell had held, and yet we'd each made the same assumption. What the hell else were we supposed to think?

Here, though, once I spotted the re-entry door on the second floor, we were able to confirm at least that the shell of the stairwell was intact, on these few floors. A couple of the guys had a closer look, and they reported up to me that there was loose rubble surrounding the door.

'Nothing we can't move, Chief,' someone shouted. From my perch one flight above, I could see that there was enough of a ceiling for two or three of these guys actually to stand without upsetting the uneasy balance of our position, and that the rubble in question was only about one foot high, so it seemed to me a prudent move, to see if we could get that door open. Who knew what we might find on the other side, where it might lead? There were two guys from Engine Co 39 on the second-floor landing – Jim McGlynn, a lieutenant, and Rob Bacon, a proby – and they set to work moving the rubble, piece by piece, careful not to upset any of the other debris. Matt Komorowski, one of Jay's guys from Ladder Co 6, scrambled down to join them and, even with his separated shoulder, managed to move some of the debris around the door. Mickey Kross, the lieutenant from Engine Co 16, helped with the light. It was slow, intense work, and we were all a little anxious about a secondary collapse, so after a few minutes of this I decided to go down to lend a hand. This was easier thought than accomplished, because even though I was just a flight of stairs above that second-floor landing, there were no real steps to take me there. There were a couple of steps to nowhere, but mostly there was about a ten-foot drop, and a sheer-face wall of debris with enough juts and jabs to it to offer only occasional purchase. I waved my light at the space between me and these guys from 39, and carved out a navigable way down in my mind. A toe-hold here, a beam to grab on to there, a short hop on to a fragment of landing. There was a landing to receive me, down where these other men had gathered. Somehow I managed to get to them without hurting myself or darkening our situation.

Everyone else remained still and silent, while we

worked the door. As our bad luck would have it, the thing was locked, so we used the Halligan to gain initial purchase and the rabbit tool to force it open – to spread it from its frame, really. The door opened in, towards the stairwell, as opposed to out, towards the second-floor offices, which was why we had to remove all that dense rubble. There wasn't much to see on the other side, once we swung the door open. There was about a foot or two of clearance, and I managed to squeeze myself partially through to peer around. What I saw on the other side wasn't worth all the sweat and effort it had taken to get there. Nothing but compacted rubble, so dense a mouse couldn't have crawled through it. I was able to step in deep enough to spot a second, closet-sized door off to the side of this re-entry door. This, I knew, housed the drain-valve for the standpipe equipment, and there was no good reason to reach for it just then but I filed the location of this water source away for later. It didn't mean much to us at that point, but who knew what kind of shape we'd be in later that day, or the following day? Could be, we were looking at hanging on in there for several days, and a water supply would be invaluable in keeping us alive long enough to await rescue. I only hoped we wouldn't have to call on it.

No one had really believed this second-storey re-entry door would lead us out of our indistinct tomb, but for a while we held on to the prospect as if it was all we had. And, in fact, it was. We had nothing. We contented ourselves with the few tools we could gather, with our still-working-but-apparently-not-transmitting radios, and with this still-standing door, thinking that together these would be our salvation, but in truth there was no way the door could have taken us anywhere but to more devastation. Really, what the hell had we been thinking? We all knew enough about building collapses to know that our prospects were not good, to know that we were likely buried beneath more than 100 floors of cement and steel and concrete rubble, and yet we seized on this door as our hope, and when this hope was dashed, our small void filled with gloom and doom. Plus, it had been something to do, some place to put our bundled-up energies, but as we pushed the door closed we each sank into our own corners of despair.

That, we all thought, was that.

And yet despite our disappointment, we seemed determined to take a proactive approach to our situation. Or, at least, I determined to do so. Firefighters are a tough, ornery bunch, under normal circumstances, and in distress we can be relentless. Here, though, that relentlessness would have to be tempered by practicality. We knew we'd have to conserve energy, restrict our movements and exercise extreme caution, else we'd never get home to our families. What the moment called for, really, was a pause, a regrouping. I didn't announce it as such, but I figured my actions would speak for me. I immediately sat back down and resumed calling 'Maydays' into my radio. A bunch of the guys had been doing the same, all through our fruitless effort on the second-floor landing, and now we were back at it – still with no good result. As I've indicated, our departmental radios had long been a source of great frustration, and this was especially true at this hour, when we couldn't raise a signal. I've since learned that the radios were working just fine, and that the reason we weren't getting a signal was because there was no one left on the scene in any kind of position to respond. Everyone was dead, or running for their own lives, or in active rescue mode. There had been no time to set up a command post, following the collapse. This makes sense, from an outside perspective, but at the time, in our silent void, it was maddening. We had no idea what things looked like on the other side. We hadn't even seen the complete devastation of the collapse of the south tower, so we couldn't begin to imagine the hell that was playing out on the plaza now that the north tower had apparently come down. To us, in our narrow view, it had to be an equipment malfunction of some kind, the fact that we couldn't raise a signal. That was the only way to explain it.

It was probably about 15 minutes from the moment the north tower had come down around us, maybe 20, and the only progress we had made was in locating some flashlights and tools, struggling with the re-entry door on the second-floor landing and determining that we'd be stuck here for the next while. I started to think long-term. I thought again of a possible water source, just outside the re-entry door. I thought of a designated space we might carve out, so that we might go to the bathroom with some measure of privacy and dignity, and at the same time not have to live with the stench of our own waste for what could run into days, or even weeks. Coniglio and Efthimiaddes were just beneath us, in a separate void, and just beneath them in another void lay Chief Prunty, so a water supply and a makeshift bathroom wouldn't have done them any good, but there were 12 of us here in this essentially open area and I needed to think what I could do for them. For us. Nobody had needed a bathroom just yet, but who knew how long we'd be trapped here, how long we could survive, how long it would take for rescuers to get to us? Who knew if rescue was even a possibility?

Thinking long-term, then, I realized we would have to conserve our resources. I instructed the men to keep their lights off, unless they were needed, and to stay off the radio. In a pinch, one light could do the work of six or seven. And, in that same pinch, one radio was all we'd need to sound our distress. Our handi-talkies worked on short-term batteries, and when they went, they went. They don't fade, they shut down, and the practice throughout the department is to charge the batteries at the beginning of every tour. Our days are carved into two tours, a nine-hour tour and a 15-hour tour, so naturally the batteries are weakest at the end of a 15-hour tour, which was just what we were facing. Remember, these guys had been dispatched to the scene just before nine o'clock this morning; the 15-hour tour had begun at six o'clock the previous night, so our batteries were pretty much at the end of their rated life expectancies. They didn't seem to be working, in the middle of all this dense rubble, and yet we couldn't give up on the prospect that we would somehow raise a signal and that this would be our connection to the outside world, if there even remained an outside world.

The plan, then, was for me to keep calling over our six channels, until I got something back. Of these six channels, only two are designated for use at any given job. There's a primary tactical channel, and if we go to a third alarm situation there'll be a command channel. For small fires, we use only one channel. And the channels change, depending on where you are in the city. Different frequencies respond better in different areas, so we always carry a card with us, indicating which channels to use in each neighbourhood. Usually, we chiefs keep the card in the glove box of our rig, and consult it on the ride over, so I knew we were working channels one and three. On this day, though, I didn't stop at the two designated channels, tactical and command. I went through all six, one by one, desperate to scare up a voice on the other end. Everyone else fell silent, lost in private thought and prayer and whatever else they were managing, as I tried to dial up a connection to the outside world. Everyone, that is, except Jay Jonas, who echoed my command to his men, to shut off their radios, but at the same time kept signalling on his own handi-talkie. I didn't know this at the time, but even if I was aware of it, I don't think I would have said anything. See, Jay was and remains a good man, and a good friend. As captain of his own company, he was in a tough spot, in terms of leadership and responsibility. To a degree, the two other lieutenants trapped in the void with us were also in difficult positions. There were two distinct companies in there, and a lieutenant from a third, and me, riding herd over all three but with no real connection to any of them. I might have grown up on the same block as some of them, gone to the same high school, helped one of them out with his first firehouse assignment, but these weren't my men. They were now my responsibility, but they weren't my men, and if Jay felt it necessary to continue talking on his radio, for whatever reason, I didn't see the point in undermining him in front of his men. He was a big boy. We were all big boys. He knew the situation, just as well as me. I put it out there, what I thought made sense, but I wasn't about to look over everyone's shoulder, making sure they were following my orders. If a light went on, somewhere in the stairwell, I didn't question it. I just figured someone needed a light, and that he'd shut it off in short order. If Jay felt the need to signal on his own radio, even though I was wheeling through the channels on mine and anxious about all our batteries going out at once, I wasn't about to call him on it. It's not my style, to watchdog people like that, after I give an order. I tell them what's expected, and if they stray from that I assume they do so with a reason. I operate on the assumption that they know what they're doing, and that they respect what I'm doing. Benefit of the doubt, all around.

Jay and I were in shouting distance of one another, but south of a yell we couldn't communicate. I couldn't hear him speaking to his men on the landing up above, and he couldn't hear me speaking to the men down by me, so there was no way for me to know he was working his radio, looking for help.

'Mayday! Mayday!' I'd holler, over and over. But I'd get back nothing.

For the next half-hour or so, things went on in this way, with mine essentially the only voice I could hear in that small, dark space. Occasionally, I'd hear Jim McGlynn, talking to his guys in the adjacent void, shouting through the floorboards, and occasionally we'd hear a bulletin from Rich Prunty over the radio, struggling to hang on, calling out for help. Other than that, though, it was mostly silent. Jay was keeping up with his transmissions, but he was keeping his voice down, probably not wanting to get anyone's hopes up until there was some rescue effort underway, and he was almost a full two flights above me, anyway, so I don't know that I would have even heard him at full voice. But they could hear me, I'll tell you that. There was nothing muted or choked or swallowed-up about my voice at this point. I was giving as good as I could. Maybe I thought if they couldn't hear me through the radio, someone could pick up on my voice through the rubble.

'Mayday! Mayday!'

Someone had dug up my bullhorn so I had that with me as well, and I was sounding the siren every once in a while, too, to see if that would bring back a response. We had our tools, we had our lights, we had my bullhorn, we had our wits, but there was no place to go, nothing to do to help our situation. Over and over, I'd go through these motions, until finally it occurred to me there was no way out, no one to hear our calls. Nothing.

Jay, meanwhile, found a voice in answer to one of his distress calls, only I didn't know he'd made contact until we had a chance to compare notes several days later. Apparently, though, he was able to reach one of the deputies running the makeshift command centre on West and Vesey streets, which was established a short while after the towers had come down. He'd reached a friend of ours named Nick Visconti, a tremendous deputy chief and a true firefighter, and they were going back and forth for a bit, trying to determine a way to get to us.

EIGHT: Contact

We'd all resumed our sitting-still, lights-out positions. We were looking at all kinds of time to kill, and no good way to kill it. Most of us had wristwatches on but nobody was checking the time. Most of us had more gear on than we needed, and it was getting kind of hot and close in there, so I imagine quite a few guys took off their turnout coats and balled them up, made themselves a little cushion to prop their heads against. We were dug in for the next while, and it made sense to keep comfortable. If I had to guess, I'd say it was about an hour and a half from the collapse, about noon, but I couldn't swear to it. We were mostly silent. The back-and-forth between Lieutenant Jim McGlynn and his trapped men had quieted. The calls and cries from Chief Richard Prunty had stopped, with the last one a gentle plea to tell his wife and kids that he loved them. And there was no communal passing of time in our own little pocket, either. There was nothing to say.

It surprises a lot of people when I tell them that, but there was very little small talk among the 12 of us trapped in that one void, and the two or three others in the adjacent pocket. I couldn't hear anybody praying, or anything like that. Nobody was talking just to hear themselves talk. And you can just forget idle chatter at a time like this. What the hell was there to talk about? We were screwed.

Up where Jay and his guys sat, though, with Josephine Harris, I imagined there were some quiet words of consoling. Jay and 6 Truck, they were on the job up there, hand-holding this poor woman. They were out of my direct earshot, and certainly out of my view, but the guys told me later they spent most of their time making this frightened, shaken Brooklyn grandmother comfortable, easing her through her fears, assuring her that help was on the way, that sort of thing. I couldn't hear any of that, two or three floors below them. All I could hear was myself on the radio, shouting 'Mayday', over and over. That, and the hideous, shrill siren from my bullhorn. When I started out, I was making these calls on the radio constantly, and sounding my siren between calls, but then I lapsed into every five minutes or so, every ten, soon enough every 15. No sense putting that constant drain on the handi-talkie batteries, I thought, if there was no one to hear it. And no sense popping our eardrums with that siren.

At one uncertain point, during this period of failed transmissions on my handi-talkie, I heard a strange call, as if it was meant to go out on the radio. It wasn't a call, really, so much as a plea, and it was coming from inside our stairwell. Loud and clear. Someone wailing about his dog. 'If anyone hears this, please take care of my dog.' I couldn't place it, at first, figured it for one of Jay's guys, from Ladder Co 6. It came from up above, in a booming voice, as if the guy couldn't help himself. 'My dog is in the basement. Somebody please look after my dog.' I can't quote the guy directly, but that was basically his message. He was all chocked up about it, real emotional, telling about his dog and how important the dog was to him, what he looked like. On and on. As he spoke, I convinced myself that it was in fact one of the firefighters from 6 Truck. I still couldn't place the voice, from our long-ago roll-call, but it came busting down those stairs like it was the most urgent plea this guy had ever made, like nothing was more important. Who the hell was it? I wondered. I was in visual contact with everyone on the second-floor landing – Kross, McGlynn, Bacon and Komorowski – and I hadn't climbed back up to my original perch after struggling with the re-entry door on two, so I figured it had to be one of Jay's guys. I'm thinking, 'The dog's in the basement, must be the basement of this guy's house, must be he's worried no one's due home for a while and the animal might starve. Or, something.'

'This dog is everything to me. Please.'

Remember, I'd just told everyone to shut off their radios, to go easy on their flashlights, to conserve whatever resources we had, and here was this guy going on and on about his dog. And what I thought was this: let the guy speak his mind, get it out of his system, give him a few seconds, don't say anything. I wasn't just issuing orders because I liked issuing orders; I was doing what needed to be done. But at the same time, I wasn't an asshole. This was an extra-ordinary situation, and there was no accounting for some of our behaviour. Let him talk himself hoarse, I thought, so long as he keeps it brief. So, I gave him a minute. Problem was, he didn't stop. He took his minute and rolled it into another. And another. He kept going on and on about his poor dog. It got to where I started thinking, 'Shit, I know more about this guy's dog than I do about my nieces and nephews.'

I'm thinking, 'Enough already. Jesus.' And, this deep into it, I'm still thinking this guy is one of ours, another fireman, talking up *our* batteries, wasting *our* resources, ignoring *my* orders, queering *our* chances of ever getting out of here.

Finally, I just burst. 'Enough with the goddamn dog,' I shouted up the stairwell. I was pissed. 'You've got guys here with wives and kids. Let's not memorialize a fucking dog!'

My response may have been a bit much, but I was past the point of giving a shit. Anyway, it shut him up. I will admit, though, that I've thought back to that moment and I sometimes come away feeling badly about it.

It turned out it wasn't one of Jay's men from Ladder Co 6, doing all this wailing about his dog, but the Port Authority police officer, David Lim, who happened to be a canine cop. I didn't know this at the time, but I learned it soon enough. The dog he was going on about wasn't his pet, stuck in the basement of his home, but his partner, trapped in whatever was left of the basement of this building. Crushed to death, was more likely, but Lim had not yet allowed himself to get to that place in his thinking. And the radio he was using wasn't one of ours, which explained how it was that I wasn't hearing him over my handi-talkie. His voice was big enough to fill that entire void, all on its own, but at least it wasn't one of our radios.

The way I shouted down this nuisance, there wasn't likely to be another, not among this group. I gave myself a couple beats to calm down, and then I was back at the radio, desperate to get through to someone. Somehow. Surely, I thought, these channels couldn't stay dead forever. Finally – thankfully! mercifully! – I got an answer to one of my distress calls. I couldn't tell you with any certainty which channel I was on, with the way I was frantically wheeling through all six, but I think it was channel three. Anyway, I got a voice coming back. I couldn't hear it clearly, but someone was acknowledging my 'Mayday', so I told whoever it was to stay on the channel. 'Just stay on!' I shouted. 'Don't leave us here.'

There was no reason to shout, the handi-talkie was right at my lips, but there was no way I wasn't shouting. After an hour and a half or so of frantic distress calls, I'm gonna be shouting, you better believe it. But I lost him. Whoever it was. However loud I shouted. I'm thinking now it might have been Nick Visconti, since he was already in radio contact with Jay Jonas, but it could have been anybody. What mattered was that the voice was there one moment, gone the next, and I kicked at the ground in frustration. Man, we were so close! I thought, 'If it has taken me an hour and a half just to get that one aborted reply, then we'll likely be here just short of forever.'

I went back at the radio, hard. 'Mayday! Mayday!' I shouted again. 'We've got people trapped here! We need help! Mayday!'

Nothing.

'Mayday! Mayday!'

Some of the guys could tell by the way I'd slightly altered my routine and kicked things up a notch that I had found someone on the other end, and for a beat or two the mood in my particular piece of stairwell grew a bit more hopeful, but then the voice was gone and we were back to whatever it was we'd been feeling, thinking. Just then, though, the most amazing thing happened. There was another voice in response, this one more distinct than the first.

It turned out to be someone I knew – pretty well, actually. Mark Ferran, a chief, out of the 12th Battalion in Harlem. Like hundreds of guys throughout the department, he was off-duty – down at the scene on his own impulse, on his own time, because he couldn't stay away. Mark Ferran didn't have to be there, but he wouldn't have been anywhere else. Most everyone I know was the same. In fact, more than 60 of the firefighters who died that day were off-duty at the time. Like Mark Ferran, they didn't *have* to be there, but they *had* to be there. And my friend Mark wasn't simply there for show. He was there to make a difference. By the time he heard from me, he'd already been all over the neighbourhood, making rescues and assists after the first collapse, and more after the second. He'd roamed as far as the ball field used by Stuyvesant High School, several blocks to the north-west, and he was now positioned across West Street, on the far side of what remained of the north footbridge leading to Three World Trade Center. The department had quickly established a makeshift command post there, on the north-west corner of West and Vesey, and that's where Mark stood when he happened to find me on channel three – not far from where Nick Visconti stood, trying to mount a rescue effort of his own in response to the occasional dispatches from Jay Jonas in the stairwell above me.

'Richie,' Chief Ferran said, once I identified myself. 'It's me, Mark. Mark Ferran. Battalion One Two.'

His voice was like a godsend. 'Mark!' I cried, and in the shouting the other firefighters figured the situation for what it was. We had that one response, and it fell away, but now we were back in business. They remained still, and quiet, but you could tell their ears were picking up a bit. That's one of the things about being trapped in a void with a bunch of firefighters; there's not much going on in terms of raw human emotion. Whatever we're feeling, we do what we can not to show it. Or, at least, that was the way of things up to and including 11 September 2001. Who the hell knows how these guys would react now, in a similar crisis? At that moment, though, there was no sense of panic, no palpable feelings of fear. And, similarly, there were no great shows of relief or hope. No highs and lows, the way you might have found if you were trapped with a group of mostly civilians, no clapping on the back or other shows of excitement at a piece of good fortune such as this. There had been no cheers from above, when Jay raised Nick Visconti on the radio, and there were no cheers down where we sat, now that I'd raised Mark Ferran. Everything was level, even, measured. I supposed this was a good thing and a bad thing, both. Good that we didn't let ourselves get all bent out of shape over each and every grim prospect or slim possibility, and bad that there was no way to muster up any shared enthusiasm, or worry, or whatever emotion could have used some mustering. Someone did flash a light on me for a brief moment, possibly to gauge my mood, or expression, but that was about it. My heart soared, but there was no good reason to celebrate – and no one to celebrate with. We weren't out of there, not by a long shot. All we'd done, really, was establish radio contact.

'Mark!' I said again. 'It's me, Richie. I got people here. North tower. The B stairwell.'

To me, this seemed like all he'd need to know to get to us. There, I thought, I'd given him our location. Now all that was left to do was wait for them to get to us. It's not like I'd have to draw him a map. North tower. B stairwell. What could be hard? To Mark, though, I might as well have told him we were hiding out in a cave in the hills of Afghanistan. 'Where the hell is that?' he wondered.

'The north tower,' I tried again. 'One World Trade Center. Take a look. You can't miss it.'

'Richie,' he said, trying to put words to the pictures I had not yet seen. 'Everything's gone.'

I thought about this, for a too-long moment. *Everything's gone.* The words just hung there, made no real sense, but I was focused enough to realize that the destruction outside was possibly worse than I'd imagined. Forget possibly. Definitely. It was way worse than I'd imagined, way beyond my imagining. Of course, this realization went hand in hand with another: the worse it was outside, the harder it would be for Mark Ferran or anyone else to get to us.

Right away, I reached for my bullhorn. 'Forget where it is,' I said over the handi-talkie to Mark. 'Put your radio down and listen for my siren.' I keyed the switch on the siren and let the thing blare. 'Mark, can you hear this?' I tried, after the peal. 'If you can't hear this, you're nowhere near us.' Then I sounded the siren again.

But he couldn't hear a thing. He told me later that over that entire rubble field, acres and acres of mass and terrifying destruction, there wasn't much to hear. Ninety or so minutes after the north tower collapsed, the place had pretty much cleared of all human life, and there was only now some sprinkling of rescue activity. There were raging fires up above, in Six World Trade Center and Five World Trade Center, and behind those buildings, to the north, Seven World Trade Center had been taken out by the collapsing rubble from the two towers, but there was surprisingly little movement elsewhere on the complex. If Mark had been anywhere near our little void, and if we had been anywhere close to the top of the rubble pile, he would have certainly heard that siren.

And so, Mark made to move. 'You've got to help me, Richie,' he said. 'You've got to give me some direction.'

I thought, 'How the hell am I supposed to give him some direction?' North tower. Stairway B. That was where my sense of direction ended.

'Ask someone where the B stairwell is supposed to be,' I instructed. 'Find a Port Authority cop, someone who knows the building.'

Obviously, I didn't need to tell Mark what to do, or how to go about doing it, but I needed to do something. I can't stress enough how helpless I was feeling, inside our stairwell tomb. There was nothing to do but bark out instructions, so I barked. Loud. And often. I kept repeating myself, offering directions I realize now were laughable. I'd suggest to Mark he could get a hold of a building plan, and measure off where the central stairwell was supposed to be, without realizing there was nothing to measure off against. Like Mark said, everything was gone, and yet here I was, trying to guide him to us along fixed points that no longer existed.

What I couldn't know was that there were mountains of rubble where the twin towers used to stand. Piles of debris a couple of hundred feet high at some points, with hundred-foot drops at other points, and beneath these piles another six or seven storeys of wreckage compacted below ground level. Imagine the undulating terrain of sand dunes in a desert, substitute crushed splinters of building materials and concrete rocks of all shapes and sizes for the sand, and you'd have some idea what Mark was dealing with. Me, I had no clue. I was going by my old frames of reference, which no longer applied. By just after noon that morning, billions of people had seen aerial or zoom footage of the scene, via television; news announcers had quickly taken to calling the area 'ground zero', but we were at ground zero of ground zero and were still in the dark, in every sense of the phrase.

The others in my area of the stairwell could sense my frustration in trying to give our location. They were all quietly hot with anticipation, waiting for something positive to come out of this sudden contact with the outside world, but nothing was happening. There was me, barking out the obvious to Mark Ferran, and Mark Ferran sending back that there was no logical way to locate the B stairwell, and that even if he knew where to look he wouldn't be able to reach us. Up above, where Jay Jonas struggled to give a location to Nick Visconti, I imagine there was much the same.

Understand, all along, while Jay was on with Nick Visconti and I was on with Mark Ferran, neither one of us could hear the other. If they came across each other on the outside, there's no way Nick or Mark would have known they were working the same rescue, talking to two guys in the same void.

After a while of this, I lay down on my back. I

realized, quite suddenly, that I was exhausted, and that I might be here a long while. Actually, exhausted doesn't quite do it; I was drained, completely spent. The feeling came over me like I needed recharging, all at once. I figured I could talk to Mark just as well on my back, so I stretched out, got closer to comfortable. I wasn't talking to Mark constantly, but every few minutes he'd be back on the radio, telling me his progress, looking for guidance. Every now and then he'd ask me to sound the siren for him, but whenever I did he couldn't hear it.

For his part, Mark was having an adventure and a half, trying to get to us. Crawling in and out of burning buildings. Dodging piles of still-falling debris. He filled me in on his efforts a couple of days later, and it truly was a monstrous ordeal we put him through. He didn't mind it – in fact, he'd lost so many brother firefighters by this point in the day he said he felt honoured to have a shot at rescuing a few. It was a long shot, but it was a shot.

First thing he did, once he had some idea of our location, was grab himself a company, much like I had done a few hours earlier in the lobby of the north tower. Ladder Co 43. It so happened that 43 was the next truck company down from Mark's house in Harlem, so he knew these guys. Lieutenant Glen Rowan, and an intact company from 43 Truck – plus a couple of guys from Engine Co 53, which shared a firehouse with Rowan's guys. They added to their mix as they moved along. By the end of the push, there were eight guys in all, in addition to Rowan, and they definitely deserve a mention here: Matt Long, John Colon, Jerry Sunden, Todd Fredrickson, Mike Regan, Frank Macchia, Jim Lanza and Tom

Corrigan. Most of these guys had been hanging around West and Vesey, waiting for some of the smoke and dust to clear, waiting for an assignment, when Mark happened by.

'You guys intact?' Mark asked Glen.

'Yeah,' Glen shot back.

Mark told them he'd just made contact with a bunch of firefighters trapped in the B stairwell of the north tower, and wondered if Glen might know how to get them there. 'You available to move?' he asked.

'I got a company, Chief,' Glen responded, 'but I've got no idea how to get to that stairwell.'

Next, Mark got on the radio with Pat McNally, the deputy in charge of the command centre, told him he had an intact company and information on some trapped firefighters, and said they were going in after us. McNally was working command with Nick Visconti, but there was no way to know that Ferran's contact with us had anything to do with Visconti's contact with Jay Jonas, so I'm guessing Pat McNally simply treated it as a separate rescue. However it played, from where Mark Ferran and company stood, it seemed impossible to penetrate on to the rubble field, because it was blocked off by Five and Six World Trade Center still raging on two sides, and apparently impenetrable mountains of debris on two more. The burning buildings were so far gone the department had given them up; the effort was now on protecting the exposures – keeping the fire from spreading to adjacent buildings.

After a beat or two spent trying to figure a way in, one of the guys from 43 Truck thought there might be a path through the upper floors of the large office building on the corner of West and Vesey, Six World

Trade Center, which housed the US Customs Office. The only problem with this suggestion was that Six World Trade Center happened to be one of those fully involved buildings, with fires so fierce it made no sense to fight them. But there was no other way.

There was a ladder leaning up against the building, placed earlier by other firefighters, and it reached to a small terrace two or three storeys above ground level that had been used as a public gathering place of some kind, so Mark told his men to start climbing. One by one, they climbed the ladder, with no clear idea where they were going, only that somehow they had to move through one of these upper floors and drop down on to the rubble field behind them. Once they got up to the terrace, they found a bunch of other firefighters working the building. Actually, they were preparing to remove the body of another firefighter in a Stokes basket – a wire basket-like stretcher used to immobilize, transport and evacuate victims, who are carried by four to six firefighters, positioned like pallbearers. As Mark took this in he thought to himself, 'Oh my God, what's this gonna be like?' There were dead bodies all over the place, and all kinds of wreckage and broken glass and chaos. But still, they pressed on. There wasn't an actual door leading into the building from the terrace, so they crawled through a window and started making their way across the floor, asking for directions as they moved. Apparently, there had been a cross-over of some kind, leading from Six to One World Trade Center, but when Mark and his guys reached it they found that it had been rendered impassable. They could no more get to us from that direction than they could from Brooklyn, so they rerouted themselves, doubling back out of Six World Trade Center, across Vesey Street, through the Verizon building on Vesey and Washington, and around the back of Seven World Trade Center, which fed on to West Broadway.

The entire south-west corner of Seven World Trade Center had been taken out by the collapse of the north tower, so our rescuers from 43 Truck were no better off there than they had been inside Six World Trade Center, but they kept moving. Mark told me later that he still had no clear visual of the rubble field at this point. There was still smoke and fire and black ash everywhere you looked, but the wind had moved things around so that here and there, now and then, you could see. The air was so thick with soot, Mark said it was like breathing flour, and without masks he and his guys had to stick their mouths and noses underneath their shirts in order to breathe – but they found ways to manage.

These were the kinds of reports I was getting on the radio. Every few minutes, Mark would think we was close and ask me to sound the siren, but obviously he couldn't hear it. There'd be no way for him to hear it, from where he was, but he didn't know that and I didn't know that. We just tried everything, hoping something might work in our favour. And if it didn't work the first time we tried it again. And again. I still had no idea what was passing between Jay Jonas and Nick Visconti (apparently they lost contact with each other for a long while). It's funny to me now, looking back, that I never bumped into their transmissions but I suppose both sets of calls were so intermittent that we didn't cross paths. The amazing thing, about all this back-and-forth on the radio, was that no one else was on the command channel. We had two five-alarm fires going, and hundreds of off-duty guys down there on their own, and on the radio there was just me and Mark Ferran. It's like the cavalry was coming and the rest of the world was put on pause. The reality was that pretty much everyone in the immediate area was dead or disorientated or busy searching for the dead and disorientated. Our ranks had been so thinned by the collapse of those buildings there was no one left to man the radios.

They were going on guesswork and vague directions, Mark and 43 Truck, and when an obstacle got in their way they moved around it, or over it, or sometimes through it. In and out of buildings. Through windows. Up and down dark stairwells, looking for an easier path. Snaking their way around roaring fires and toppled buildings. On West Broadway, they came across a large group of firefighters, and a tremendous debris pile. Biggest one they'd come across yet. There was no way to cross it, or climb over it, and in angling around it Mark ran into a firefighter from 18 Truck who seemed to know a way to get to us. How this came up, I have no idea, but Mark made it happen. Everywhere he went, he stopped people and asked them if they could lead his guys to where the B stairwell used to be, and this one firefighter seemed to have an idea. He told Mark to head towards Church Street, behind Five World Trade Center, which was also roaring pretty good by this point. But Mark wanted more than directions. He'd had enough with directions. He grabbed the guy from 18 Truck, and told him he had to come with them. Literally grabbed him, by the back of his jacket.

'But I'm with my company,' the firefighter said.

'I got people trapped,' Mark said, insisting. 'You gotta show us how to get there.'

Mark can be a persuasive sonofabitch in a moment of crisis, and he managed to pull this firefighter from his company to take him and the guys from 43 to the heart of the rubble field. This firefighter knew these buildings, so Mark wasn't letting him out of his sight. Together they moved through a bookstore at the south-west corner of Vesey and Church. Borders Books. I'd been to that particular store a bunch of times, and when Mark retraced his steps for me I could picture it clearly. The place was dark, save for the exterior light that managed to pierce the grey film of soot that seemed to hang everywhere. The electricity was history. There were big picture windows rimming the store on the street, but the interior was so vast that by the time they reached the middle of the store there was very little ambient light. They raced through the bookstore, past rows of toppled bookshelves and scattered books, down a non-working escalator, on to the concourse level of Five World Trade Center. Supposedly, there was another escalator that fed directly on to the main plaza of the World Trade Center complex, right into the north tower. The firefighter from 18 Truck moved with such purpose Mark felt sure they'd be on to that field in no time, but they brick-walled where another bank of escalators used to be. There was a giant hole on the concourse level of that building – and once again, no clear way to get past it.

'Richie,' he called to me over the radio, at what he and I later figured to be this particular impasse. 'We still can't find a way in to you, but we're coming. We're coming. Hang on, 'cause we're coming.'

I thought, 'Of course we'll hang on. What the hell else are we gonna do?'

NINE: Light

None of us knew how much time had passed. It could have been another hour from the moment I'd raised Mark Ferran on the radio. It could have been another half-hour. To tell the truth, I wasn't thinking much about the clock. Hell, there was one on my wrist, but I wasn't going looking. It was getting late, and that was all that mattered. Late to expect an easy rescue or escape. Late to go back to how things were. Late.

We had already burned precious resources following our false hopes. The push on the second-floor landing, to dig out that re-entry door. The search for tools. The drain on our lights, and our handi-talkies. The emotional toll of listening to one of our brothers – a chief, no less – quietly expire in an adjacent void. The holding up against the kind of tension that could snap a bridge. The endless waiting on the next miracle. We were here on one miracle; we'd get out on another. How it would find us, I had no idea.

Suddenly, I felt tired. Completely, overwhelmingly tired. All over, and all at once. Like it was all I could do to support the weight of my own eyelids. We were between transmissions on the radio, Mark had just run into that hole beneath Borders Books where the escalator used to be, and our tiny stairwell tomb was deathly silent. Everyone in earshot had been listening in on our transmissions, almost like it was a radio broadcast of some kind, but no one was saying anything. That's something else people keep commenting on when I share my story, the way we all kept to ourselves for those long, difficult moments. We were connected by where we were, and what we were hearing on the radio, but we kept conversation to a minimum, like it too was a resource we thought we might need to keep in reserve. I couldn't see, but I pretty much knew where everyone was by the sounds of their breathing, by where we had pictured each other at last light. Everyone around me on that second-floor landing, anyway. I still hadn't laid eyes on Jay Jonas and the rest of his company, a couple of floors above. And I still hadn't seen Josephine Harris, the 59-year-old grandmother.

There was an overlong gap in the transmissions from Mark Ferran, and I fell into it like I had nothing left. I imagined from the quiet and stillness that other people were also closing their eyes to this nightmare. Maybe a few folks had nodded off. Lord knows, I was feeling drowsy, lethargic. Like I'd had enough. There was still plenty of fight in me, don't get me wrong, but I'd had enough, and I figured the thing to do was shut down for a beat or two. Gather my thoughts, my resolve. Figure a plan. The fight could wait for later. Either it would be there, or it wouldn't. Either I'd need it, or I wouldn't.

As I lay there, on my back, my eyes momentarily closed, I started to think I could drift off, and then my mind raced to how it was that I was so quick to want to drift off. In my light-headedness, I thought maybe we were being poisoned by carbon monoxide fumes, or a lack of oxygen. Some unnatural element, sent to knock us out and finish us off. I wasn't fighting sleep so much as I was confused by it. I thought again of Debbie and the kids. I thought of those two firefighters from 39 in the pocket down below, Coniglio and Efthimiaddes, who'd made it down to the lobby before doubling back to the stairwell for their lieutenant, Jim McGlynn, who lay alongside me here on this second-floor landing, and how it was the doubling back that saved them. If they had continued out, they'd have been caught in the crush.

I thought of the Battalion Chief Richard Prunty, two voids below. He'd been seriously injured in the collapse, his voice weak on the radio, his body no doubt in tons of bad shape. I thought how I hadn't heard from him in a while. All along, that first half-hour or so, he'd been calling for help. I'd run into his transmission on my frantic search for an outside signal and had told him to stay calm, that help was on its way. Jim McGlynn spent some time talking him through as well. The Chief's voice faded a bit each time, but we kept telling him to hang on. We told it to him so often I started to believe it myself, and now that it was finally true – help *was* on its way – it rang hollow. Yes, Mark Ferran had gathered an entire company to come looking for us, and yes, they were determined to get to us as quickly as possible, but they weren't making any progress because progress was impossible. I imagined that Prunty was gone, along with his voice and my empty promises, and that we would be gone soon, too.

That's the way this sudden weariness had me thinking, that I would just drift off into sleep and never wake up, and as it occurred to me I thought it wasn't such a bad way to go. Given the circumstances, given what could have happened, what could happen still, it wasn't so bad. Kind of peaceful. Shit, it beat starving to death. Or slowly suffocating. Or dying of thirst. Or a secondary collapse, or fire, or any of a dozen other terrors that had surfaced in my fevered thoughts.

For some weird reason, I thought of the words to that old Simon and Garfunkel song, 'The Sound of Silence'. 'Hello darkness, my old friend.' It's funny, the way certain things hit you, at certain times. It just came to me, this line, and underneath it was the sweet voice from the record that had been burned into my growing-up, like background music, and it felt to me like I was embracing the quiet, the stillness, the darkness. I heard it in my head, over and over, like I was my own radio. 'Hello darkness, my old friend.'

The song, the silence, the enveloping stillness... it was all weirdly soothing. No, it wouldn't be such a bad way to go, to just drift off like this. So I gave myself over to it. Sleep. Darkness. Death. Peace. Whatever it was, I was ready, and as I gave myself over to it I was washed up in waves of memory. Once again, it was like it was in the movies. Like people talk about, back from near-death experiences. The whole of my life flashing before me, all at once, down to every detail. Everything mixed together in this warped collage. Pictures, conversations, moments... all right there for the figuring. I'd caught some of these same scenes earlier, the moment the building began to shake and topple, but now I settled in for a second showing. I had all these clear pictures and recollections in my head, vying for focus.

There was my first job, as a green proby – or 'Johnny', in fire department terms – fighting a tenement fire in Spanish Harlem like I knew what the hell I was doing. Like I mattered, finally. I could see myself, clear as yesterday, struck by the incredible strength and single-mindedness of the good men around me, awed by the responsibility of the path I had chosen. I remembered wanting to get out, soon as I got in. The heat of that first fire was so intense, the smoke so thick and heavy, I almost couldn't take it. But the guys in my company kept going, and I was pulled in with them, fixed to the same length of hose. We went in as far as hot, and then as far as too hot, and then as far as too damn hot. And then we went in some more. It was incredible. You get used to it after a while, but that first pass was tough. And it wasn't just the intense heat that had taken me by surprise. It was everything that came with it. It was one thing to fight a simulated fire, and quite another to battle a blaze where lives were at stake. To be on the line when something was *on the line* was a tremendous burden and a tremendous blessing, both, and I held this vision of my rookie self, with the dull orange proby badge on my helmet, for the longest time. And as I held it, I allowed myself the thought that I'd done good. Come a long way. Saved a lot of people. Made a difference.

This vision finally gave way to another: me, still a proby, desperate for the respect of my brother firefighters. Me, peeling potatoes in the kitchen. Me, scrubbing the toilets and mopping the floors. Me, asking in. Engine Co 91, on 111th Street. A couple of months into my tour and I still hadn't held the knob – the nozzle at the fighting end of the hose. That's a big deal, for a rookie firefighter, first time he gets to hold the knob. Oh, I'd held it – to wash up after a job, to place it back on the rig – but I'd never *held* it, fighting a fire. It was a position of great honour, to be holding the nozzle, and I went to job after job wondering when the honour might fall to me, if it ever would. There were some guys, I knew, never got to hold it.

Then, there was another tenement fire, also in Spanish Harlem. Jim McCloskey was senior man. The knob was his honour, and he never gave it up. All these years later, I still thought of Jim McCloskey as the most amazing firefighter I'd ever known. Tough. Unyielding. Fearless. A guy who led by deed, by the way he carried himself, by the way he battled. Work hard, play hard… that was Jimmy. Al Quinn, another great firefighter, grabbed the line and offered the knob to Jimmy, and as he did, something came over me. I don't know what it was, or what it meant, but there was no stopping it. I stepped between Al and Jimmy and said, 'I'll take it.' Me, a snot-nosed, pissant Johnny, reaching where I didn't belong, not yet. But for some reason Jimmy didn't laugh at me, or shoot me down or brush me off. Our eyes met. He flashed me a good, long stare, and in that gaze he saw my future. Or, maybe, he saw his past. Jimmy never gave up the knob, but he gave it up to me on that day. Just handed it over, like I had it coming, and then he talked and pushed and cajoled me through the door and up to the fire. It was my rite of passage, my trial by fire. I was a kid, being set loose on a two-wheeler, first time without training wheels. Back in the days of iron men and leather lungs. Our war years. The early 1970s. Two, three, four jobs a tour. Big jobs all. Landlords torching their own buildings for the insurance. Tenant arsonists looking to move up on the city's subsidized housing lists. Developers angling for a better deal. One fire bigger than the last, and another one always waiting. *I'll see you at the big one*. That was our rallying cry, our sign-off, and there was always something bigger. We lived for something bigger, something more. We had our masks, but the custom was to leave them in their cases on the rig. We saved them for the tough jobs, only the tough jobs never came. We were too tough to let them. And there I was, green, holding the knob, thrilling to the adrenalin rush, grabbing on with everything I had. Man, the sheer force of all that water! The pressure! It was all I could do to hold on, but I held on, and eventually we put that sucker out, and I could still see my kid self, spilling from that shell of building, filthy, drained, sweating. I had arrived, and everyone else knew it.

There are no words, really, to explain the rush of that moment, first time around. The elation. The immediate gratification. The power. The purpose. All of it magnified if you've been fighting a fire in an occupied building, saving lives. To think that you actually put out the fire. You. The nozzle man. Everyone else, from chief on down, just a supporting player to your central role. It's a feeling that's second to none in the whole world. Better than sex. Better than hitting the winning grand slam in the bottom of the ninth inning. And the first time is always the sweetest.

From there, it was a straight line to everything else. My parents, still in Staten Island, same house we lived in when I was in grammar school. My father a city sanitation worker, my mother a nurse. Good jobs, both, but not enough to pull us from the projects until a triple at the track up in Monticello left us with a down payment and a story for a life-time. They'd be watching this drama on the television, I felt sure, worrying where I was, why they hadn't heard from me. My daughter Lisa, too, would be caught up in it. She was at school, Pace University, just a couple of blocks away, and she'd be convinced that I was inside one of these buildings, and she'd be right. Stephen would be at high school in New Jersey, far enough away from these events to push them from his mind. And Debbie... she was always a little bit nervous about the job, but not in any kind of front and centre way. Always, she found ways to talk herself down from her concerns, to distract herself, and she'd find one today. She'd be at the hospital, working. She'd be okay, probably thinking I was still at the firehouse, still a couple of miles from the unfolding disaster. Not yet waiting on word, because there was no word to wait on – going about her day as she knew I'd be going about mine.

Some time later, she told me what she was actually doing, during these tense, uncertain moments. She was assigned to the newborn nursery that morning, catching glimpses of the devastation on the news, holding tight to someone else's baby as if it were her own, thinking of Lisa two blocks away. Thinking of me, refusing to believe I was anywhere near the scene and at the same time refusing to believe I was anywhere else.

Debbie... I drifted back to how we met. We were at Staten Island Community College at the same time, but we didn't meet there. We had to go to the Jersey shore to cross paths, and once we did there was no uncrossing them. I'd just graduated from Baruch College, and went directly into the Police Academy. I thought law enforcement would be my career – and it was, for a time. I was a street cop on the lower east side, and by the time I was laid off, in the financial freeze that hit the city two and a half years later, I was bored. What I really wanted, I knew, was to be a firefighter. I'd had a paper route as a kid, and I used to deliver to the local firehouse, and I remembered feeling the pull of the place even then. These were great, good guys. I wanted to be one of these great, good guys. I'd ride the pole, or hang out in the kitchen. I grew up in a blue-collar, civil service neighbourhood; I was meant for this kind of work. It's not like I grew up wanting to be a fireman my whole life – hell, I wouldn't have landed in the police department if that was the case – but when the call came, when the city started hiring again and the department activated the old list of people who had taken the test, I grabbed at the first opportunity. And, in the grabbing, I realized that this was where I belonged. And, also in the grabbing, I managed to hold and hang on to Debbie, who for some reason agreed to throw in with me. It was about five years after we met that first summer after college. The joke is that it took me that long to convince her I was the guy for her; the truth is it was me who needed the convincing. I guess I needed to find my footing on a job before I could make a life with someone else. But the fire department had quickly become home, and soon enough I realized Debbie had become home – and the rest of my life just spread out before me. Home sweet home.

These drifting thoughts were interrupted by another wailing, from inside our stairwell. Same voice as before. David Lim, the Port Authority cop, moved now from concern for his dog to the conviction that he was being soaked with jet fuel. His cries pierced through my fading consciousness and shook me alert. Slowly, at first, as if the interruption was part of whatever visions and recollections and runaway thoughts had been playing in my head, but I came clear soon enough. This poor guy was a little panicked, I realized. I could understand that. It was getting weirdly claustrophobic and dreamlike and terrifying, where we were, and it was a wonder to me it took this long for one of us to crack. If I could even call it cracking. Maybe he really was smelling jet fuel, for all I knew. Even from my uninformed perspective, I guessed there had to have been thousands of gallons of fuel pumped somewhere into that building when the plane hit. Maybe it was dripping on him. He was about a flight above me, so I couldn't tell what he was experiencing. I didn't smell anything, though. I've got a fairly honed sense of smell. My nose can distinguish between a wood-burning fire, or an electrical fire, or burning plastic or garbage. There are a lot of different smells to smoke, a lot of different smells to fuel, and here I didn't smell a thing. Just ash and concrete dust, and the low-hanging sweat and effort of a bunch of beefy, overdressed firefighters in too-close, too-hairy quarters. But Lim kept going on and on about it, to where he had a few other guys whipped into this heightened state of agitation, and the dark silence of our stairwell tomb was fully broken by shouts of, 'I don't smell anything, do you?' Or, 'There, right there, it's dripping... I can hear it!' It was the one piece of back-and-forth conversation we'd had in over an hour, and it was about the last thing we needed, to get all bent over this phantom fuel.

I worked hard not to get caught in the same swirl, but at the same time my mind lit on to what it knew. If there was fuel – and, surely, there *had* to be fuel, somewhere – the vapours could combust in a flash-point and we'd be toast. The fuel itself doesn't actually burn, but the vapours surrounding it are a torch waiting to happen. Once it hits its ignition temperature, we're gone, and the fact that I wasn't smelling anything now didn't mean I wouldn't be smelling anything soon, didn't mean David Lim wasn't smelling anything one flight above me. But, then, we all had enough to worry about without worrying about this.

Soon, Lim quieted, and the other guys all satisfied themselves that we were in no immediate danger. We returned to our private, rambling thoughts. Mark Ferran was out there somewhere, looking for a way in to the rubble field, but I hadn't heard from him in a good long while, and I didn't want to pull him from what he was doing to answer my call. He was making his progress, going forward, doubling back. I figured he knew where to find me if he had something to report, so I didn't see the point in reaching out to him on the radio. I just lay back down and closed my eyes again and let the black silence do its job. A few minutes of this, and I was back into the same lethargic state I'd left just a few beats earlier, before the contagious panic over the fuel. I was drifting off again, unable to fight it. Unwilling to fight it, really. My eyes were closed, and then opened, and then closed. It made no difference. Actually, it made a difference in terms of comfort, because, having been filled with dust and smoke, my eyes were now stinging something fierce. Man, it killed! I thought about returning to that water source, outside the re-entry door, to maybe flush them out a little bit, but dismissed the idea as too much trouble. I didn't want to set off a whole new wave of panic, me scrambling for water, so I tried opening and closing my eyes, wondering if I'd be more comfortable, one way or another.

Mostly, though, my eyes were closed. My mind was open, but my eyes were closed, and I returned to thoughts about my family, about my circumstance, about the miracle that might never find us beneath all this rubble. Acres and acres of ruin. Mountains of concrete and iron and dust. For some reason, I started thinking about a two-storey building collapse out in Queens, from just a couple of months before. The Father's Day Fire, that's how it became known in the papers. It happened on Father's Day, and three firefighters, all young fathers, had been trapped in the rubble. Three guys, with just two storeys crumbled down around them, and the rescuers still couldn't get to them in time to save them. Two storeys! And here we were, beneath the rubble of more than one hundred storeys! If there was anything I didn't need to get my mind around at just that moment, it was the fate of those poor fathers in Queens, the fate that surely awaited us. I worked it out. Two storeys, one hundred storeys. Our situation was 50 times worse.

Please, God, make it quick.

So much for the power of prayer.

I willed myself to think of anything but that Queens fire, anything but where we were. Or nothing at all. To go completely blank, empty… that was the thing. Take myself away from this dismal place. Soar on to some other plane, some other way of thinking. But I didn't have the tools to free myself from my fears. I was stuck – here in this stairwell, and here on this train of thought. About the best I could do was realight on the money situation that would face Debbie and the kids after I was gone. I reminded myself they'd be taken care of, and as I thought of this, it left me feeling calm, and assured. For some reason, I landed on the phrase 'rest assured', and I took it as permission to do just that. To rest, assured. Of course, my kids would go the balance of their growing-up without a father, without a loving father, and there was nothing good about that, and my wife would be a widow, growing old alone, probably never knowing what had happened to me, what these last moments had been like, but I allowed myself to take comfort in the money part. In knowing they wouldn't have to worry about the house, or the bills, or tuition.

It felt to me as if I was drifting in this way for the longest time, from loose thought to loose thought, fragment of memory to fragment of memory, to where I might have been hallucinating, because at some point I looked up and thought I could see a flicker of light, about three or four storeys above me. It wasn't even a light at first, but a brightening. Better, like the very last coal to fade from a fire. Like a pinprick in the black sky that closed us off from the rest of the world. It didn't make any sense to me, this light, the way it appeared out of nowhere. It appeared, and then it was gone, and then it was back again, and really, it was a total aberration. We'd been marooned in here for the longest time – surely, one of us would have noticed the light by now. But there it was. Either it was a hallucination, or it was the real deal, so I shouted up to Jay Jonas, to see if he could make it out too.

'Hey, Jay,' I called. 'You see that light? Straight up above. Am I imagining things or what?'

I held my eyes on that pinprick as I spoke, and the damn thing seemed to be getting bigger, and bigger, and bigger. It went from a pinprick to a fist to a basketball to a genuine hole in the sky. All in the time it took to describe it and point it out to Jay.

'Jay?' I tried again.

There was a long pause. Finally, Jay spoke. 'Yeah,' he shot back. 'I see it. What the hell is that?'

We had no idea. Whatever it was, it was growing brighter. Wider. Clearer. Within just a few minutes, it seemed, there emerged this wondrous opening at the top of our small cave. A hole in our man-made sky. It was as if our very will had punched through to the other side. There was no explaining it, even if an explanation would find us sometime later. What had happened was, after the collapse, there was no place for all that thick, black smoke to go. The hole at the top of our void had been there all along; we just couldn't see it. It was like a small chimney, but the smoke seemed to settle in such a way that it took hours for it to clear. There were no extreme temperatures down where we were, nothing to chase the smoke and ash and dust up towards that flue in any kind of hurry, so it just kinda hung there. For the longest time, it hung there, recirculating. On top of that, it turned out that the buildings next to us, Five World Trade Center and Six World Trade Center, were on fire, raging, and the residual smoke on the outside was so intense it left another huge cloud atop the small funnel into which we had fallen. We've all by now seen pictures of ground zero in the hours just after impact, of all those people absolutely covered in dust. We've heard reports of people blocks away being unable to see their hands in front of their faces, so that's how dense the smoke was, and it remained so for hours. Here, at the eye of ground zero, the effect of all that smoke was compounded. That, and all the dust and ash, had effectively obscured this opening from view. Until gradually the cloud lifted inside our stairwell, and the winds shifted outside our stairwell, and we could see hope where there had been no hope at all.

We were all staring at it, by this point. If I was hallucinating, it was a shared hallucination. And as it cleared, I thought I saw a piece of sky. Could it be? I wondered. Could it possibly be? And then I saw it again, and this time there was no mistaking it. I thought, 'Holy shit!' I saw a patch of blue. Remember, it was a vividly bright day outside, and here was a brilliant patch of blue.

I immediately got on the radio to Mark Ferran. 'Mark,' I said, 'Mark, I can see the sky! I can see the sky!'

'What?' he sent back. After that disappointment below the bookstore, he had taken three guys from 43 – John Colon, Todd Fredrickson and Frank Macchia – and splintered off from Glen Rowan and the rest of his company, thinking they could cover more ground as two units. He wasn't sure he'd heard me right. After so long in darkness, how was it possible that I could suddenly see the sky?

'The fuckin' sky, Mark!' I said again. 'Blue sky.'

'Hey,' he said back. 'All right. Now we've got something.'

Yes, we had. Right away, the mood all around me began to pick up. There was no jubilation or celebration or anything like that, but the general weariness and hopelessness were gone. Our situation brightened, and our mood along with it. No one said anything but we could all sense the shift. This was the miracle we'd been waiting on.

My first thought was to shine a beam of flashlight towards the opening, which I figured was about 40 to 50 feet above me. I thought maybe Mark could catch the light as it poked through the rubble, but of course he couldn't. It had been dark inside our cave, but outside it was bright, full-frontal daylight. It was a stupid move – I realized this even as I went through the motions – but I did it anyway. No way Mark could see it, middle of the day, but he was good enough to look for it, and report back to me with his disappointment. But even as the others sagged at the report, I realized that, six or seven hours from now, it would be dark. At that point, there'd be no way he could miss it. We were as good as found.

Next, just to speed things up a bit, I shot a bull-horn siren call towards the opening, but Mark couldn't hear that either. I still thought he'd find us soon enough. We could all hold on another couple of hours, waiting on nightfall. We had water. We were all in relatively good shape. Banged up, but in good shape. Even Josephine Harris was hanging tough, from the reports I was getting from above. We'd lost Richard Prunty, I feared, but the rest of us would make it out in one piece. We'd find a way to get to Jeff Coniglio and Jim Efthimiaddes beneath the rubble where we lay, and we'd make it out of here. We'd survive the collapse of the tallest friggin' building on the planet, and make it out in one piece. In 14 different versions of one piece.

How about that? I marvelled – and somewhere in the marvelling I decided I couldn't wait for nightfall. I was too jumpy, too pumped, too desperate to get out of there. Plus, a part of me thought that hole could close up on us as magically as it had opened. I thought the time to move was now.

'Can you get to that hole, Jay?' I shouted up. 'Whatever it is, can you get to it?' Jay was closer to the light than I was, so I wanted his perspective.

'I don't think I can,' he said. The wreckage was too unstable, I guessed. There was no purchase. There was still the very real and very likely possibility of a secondary collapse. We had been gifted these extra moments inside this stairwell, and they could be taken from us at any time.

But I kept staring at the light, wondering how to get to it, what was on the other side, where it might take us. And as I stared, I determined to reach it. 'I think I can get to it,' I said, for no good reason. Honestly, I had no idea if I could get to it, but it was something to say. Something to shoot for.

TEN: Escape

I set off for the light at about two o'clock in the
afternoon, best I could figure. The timeline didn't
matter to me then, but it feels important to me now,
in the retelling. All we cared about, in terms of time,
was that there'd be enough of it to see us through. As
I've said all along, it never occurred to any of us to
look at a watch, not even to Mark Ferran on the
outside, but it had to have been at least three hours
from the time the building came down to the time
that bright hole opened up in our black sky, probably
longer. I'm thinking, an hour and a half or so until I
raised Mark on the radio, and another hour and a
half while he tried in vain to reach the rubble field,
and somewhere in the balance another chunk of time
to account for the agonizingly slow tick of the clock
at such a tense moment. God, for all I know, it could
have been four or five hours, all told. For all my
asking, there's no one in the department in a position
to put a clock on what we went through.

Whatever it was, however long it took for that
bright hole of sunlight to appear, it was a good long
while to be trapped in such a vacuum. Remember, we
had no idea what was going on outside our tomb of
a stairwell. What the whole world was watching, we
couldn't begin to comprehend. We couldn't even
make an intelligent guess at it. All we knew was what
we could piece together from various radio reports,

and even here we were sometimes going on unsub-
stantiated information. Some of us, I later learned,
didn't even realize that the south tower had come
down at 9:59 that morning, so their thoughts about
our position weren't necessarily as bleak as mine. I
thought we were lying beneath 110 storeys of steel
and concrete, destined for a legendarily slow death;
they thought we had merely suffered an interior
collapse of some kind, involving just a few floors of
the building, that the building itself was still standing,
and that we were primed for rescue.

As it turned out, we didn't have the first clue.

What had happened was that the core of the building on these lower floors had somehow held, while
the exterior portions had run out of places to go.
That's the nutshell. Beyond that, there was no
explaining it, and yet when I've talked to people since
the collapse, they tell me it makes a strange kind of
sense – it was basic physics. Who knows, maybe it
was. Yes, 110 storeys had crumbled, but the debris
seemed to have fallen *around* us, or been pummelled
into the ground and sub-basement areas surrounding
us, more than it had fallen on top of us. It's kinda like
the outside of the tower melted away, and the debris
was so solidly compacted at the core that it somehow
pressed against itself and a pocket was allowed to
form in its middle. Or, something. Of course, the
wreckage was down and around and on top and
all over, strewn all over the damn place, but when
the dust settled we were somewhat near the top of
the left-behind pile of twisted steel and concrete.
Consider that the World Trade Center complex had
taken up about 16 acres of real estate, but that the
rubble field, after both towers had collapsed, had

spread to an area covering as much as 50 acres. Streets four, five, six blocks away were choked with heaps of debris; abutting buildings were taken down by falling blocks of concrete and enormous beams; and we remained in the carcass of the north tower, at or near its base, only 40 or 50 feet from blue sky.

From the outside, Mark Ferran told me later, there appeared this one lonely wall, reaching up from the rubble. It must have reached about ten storeys high, from the top of the pile – and it was the only part of the building that still looked like a building. Everything else was like a pile of pick-up sticks. That's the way Mark described it, and it's pretty much on target. It's like everything was standing and solid, and then let loose to drop and scatter, except for this one wall, and inside that wall was the shell of our stairwell. Ten storeys high seems about right, from my interior perspective. We were down on the second floor, 6 Truck was up on five, and this dawning light seemed to be another 20 or 30 feet above them.

Jay Jonas and his men had explored the portion of stairwell where they had been sitting some time earlier, and determined that the stairs themselves were somewhat shaky. They weren't really attached to anything; it's like they hung there on sense-memory. I could certainly see being freaked about climbing on them if you had no particular place to go, but once that light presented itself we had every place to go, so I was gone. I shot up those couple of flights to Jay, mostly taking these untethered stairs but also reaching for whatever toe- or hand-holds I could find and claim along the side walls. There were all kinds of protrusions, juts and jabs reaching into our air space, and these presented themselves as key climbing purchases, so that when the staircase gave out for a stretch I could shimmy up to the next batch of steps. I saw them not as obstacles, but as climbing tools. Compacted girders, and metal beams, and I've got no idea what else, just hanging there, feeling solid enough to grab on to for support, so that I might lift myself another few inches. Amazingly, nothing was hot to the touch. Many of us have since read that temperatures in the jet-fuelled fireball eventually reached somewhere between 1,500 and 2,000 degrees Fahrenheit, which was what had softened and compromised the steel supports of the tower structure, but that killing heat hadn't reached our void, so I was free to grab and clutch and climb without worrying about burning myself. Actually, it didn't even occur to me to think of the heat; I wasn't thinking much of anything, just moving. If I stopped to think about what I was doing, I would have been scared shitless, with the tentative hold I had at each moment, but who the hell had time to stop and think? I was going for it, all out, all at once, without a thought in my worn-out brain but to reach that damn light. Whatever it took.

And it didn't take much, thank God for that. Or, I should say, it didn't take long. I bounded up to Jay in the time it might have taken to rethink my plan, and I sat with him a while, went over a couple of things, wondered how it was we were still alive, considered our course from here. This was the first time Jay and I had actually made physical contact with each other in the three or four hours we'd been trapped, and he was a God-welcome sight. To see someone you know, at the hopeful end of such a dark moment... man, it's a real release. And a relief. Here's one thing I learned that fateful day: it's better to die with a friend than a stranger. And here's another: it's better to fight and survive with a friend than to die with one.

Jay filled me in on what had been going on with Josephine Harris, on how he and his men had been taking turns sitting with her, comforting her, talking her through her fears, holding her hand. She'd been holding up fairly well, all things considered – and here's an instance for which that phrase was invented. *All things considered.* Every last thing, and then a few more besides, and this tough, brave woman was stoic and composed, and finding reserves of strength and resolve she likely never knew she had.

Talking with Jay also helped me to see the void from his perspective, which I quickly learned was a whole lot different than mine. There was far more headroom to work with up here, two to three flights above where I had passed most of my time, after dropping down to that second-storey landing to work that re-entry door in the first moments of our incarceration. They also had a bunch more elbow room, too – a full width of stairwell, whereas we had been squeezed at each side, our steps collapsed inward in such a way that all that was left were slivers of concave slabs. Compared to our accommodations down below, this was the friggin' Ritz – but then, compared to the Ritz, Jay's spot was pretty much a hell hole, same as ours.

Now, here's where the difference in perspective kicked in. To my mind, there was but one option: climb up and out. Jay, I quickly learned, had other ideas. He was real keyed by an elevator shaft that had somehow alighted in his view, and he talked to me for a bit about scrabbling over to it and making our way down. He'd already checked it out with some of his guys, by flashlight, and once the blooming light above revealed the shaft for what it was, this emerged as his plan. It made no sense to me, but to Jay it seemed clear; I guess he was thinking in terms of the building we had entered, and that we still had to find our way *down* from the fourth or fifth floor.

I saw our situation differently. 'Jay,' I said, 'we're not going down, we're going up.' I indicated the opening in the stairwell ceiling, which seemed to be about 40 feet above where we sat. 'That's our way out of here,' I said.

'How you gonna get to it?' he asked.

'Don't worry,' I said. 'I'll get to it.' This close to open air, there was no way I wasn't trying.

Before I lit out for that opening, I stopped to talk to Josephine Harris, see for myself how she was doing, and like Jay had said she was doing surprisingly well. She smiled weakly, and I touched her shoulder gently and was gone. Up, and gone.

I was still dressed in my turnout gear, despite the heat and closeness of that stairwell. I still held a flashlight in one hand and my bullhorn in another. My helmet was gone. I'd never recovered it after that first fall. I had a pair of work gloves in my belt, which I thought I might need in the climbing, but at this point I was still bare-handed, and hoping I wouldn't tear those mitts open on some jagged piece of something. The going was pretty easy at first, the stairs held just fine and reached almost to the source of the light. It wasn't a straight shot, those steps, but they filled most of the space between where I was and where I was going. Here and there, I'd have to find a less certain path, if the staircase gave out momentarily,

as it did on one of the return landings, and I'd find
something to grab on to, and hoist myself up another
few feet until the steps continued. I didn't need the
flashlight to light my way at this point, but I didn't
want to lose it just yet, so my fingers were slipped
into the handle, which kept my hands otherwise free.
As I climbed, I marvelled that the beams of exterior
light piercing our void were bright enough to catch
the little dust particles hanging there. If I'd had a
camera, it would have made a stirring shot, the way
the ash and soot formed these dancing patterns in the
bright sunlight – kinda like inching up to the rafters
of a dilapidated barn, and watching the sun bake
through what's left of the roof. As I reached closer to
the source of the light a part of me felt I was being
pulled.

Like I had no place else to go.

Below me, whenever I looked down, I could see the
hopeful gaze of Jay Jonas and his men, and Josephine
Harris, stealing glances skyward, measuring my
progress, careful not to let their hearts get ahead of
them. I don't think I looked down more than once or
twice, but the image held. Matt Komorowski and
Mickey Kross and those guys from 39 were too far
below for me to spot, but I'm sure they were getting
progress reports from 6 Truck – and below them
were those final two firefighters from 39, Coniglio
and Efthimiaddes, stuck in their separate void, no
doubt plugged in to what was happening through the
other guys in their unit. And it's not like it took a
whole lot of time, to get from the uppermost spot of
our stairwell to the hole at the top. Four or five
minutes, at the most, and the only reason it took *that*
long was to account for the stops and starts I needed

to make along the way. To figure a good route to take. Like I said, it was no cakewalk, but I made quick work of it, I'll tell you that.

What I found, when the stairs gave out and the hole opened up, was everything. Intense bright light. Utter devastation. No signs of life, or movement. No noise but the rip of wind. Naturally, this was the first I was seeing of the unfathomable, mind-boggling destruction on that plaza that morning. To burst on to such a scene, essentially unaware, was to question your place on this earth. I had no frame of reference for a sight like this, no way to prepare for it. You have to realize, most everyone else on the planet had seen these buildings collapse before looking on that massive rubble field, which I'm thinking must have pre-conditioned people to take in this scene, but I had no idea. Absolutely no idea. And as I looked, it just about knocked the wind out of me, to think the few of us had apparently survived such complete catastrophe. I say *apparently* because at just this moment, as I was taking in this unimaginable scene, it occurred to me that there was no good way down and out from where we stood. Fires were raging all around, in every building I could see on the perimeter of the Trade Center complex.

What had seemed a hole in the ceiling from my perch on the second-floor landing was actually a hole in the side of the stairwell, opening up to a kind of balcony, looking down on to the rubble field. Remember those photos from the Oklahoma City bombing, where the front face of the building was torn off and you could see directly into all those rooms and offices? That was the effect here. Like the stairwell had been sliced down the side, the exterior walls of the building ripped from the frame, and this one landing had been left exposed.

I couldn't get over the intense sunlight. I noticed this first, and most of all. It was the most incredible thing. There were still tons of black smoke and heavy cement dust that had yet to come to rest, but there were pockets of clean air in the shifting wind, and every here and there I'd get a face-full of clear blue sky and bright sun. The sun was so damn bright I had to close my eyes to it, and for the first time I started to realize what kind of damage had been done to my eyes from all the ash and smoke in that stairwell. It killed to look at anything – this piercing, stinging, *astonishing* pain – but at the same time it killed not to look. My eyes were a little more comfortable if I kept them closed, but there was no closing them to what I was seeing. After all those dark hours, thinking all those dark thoughts, I couldn't look away from this. I was facing east, towards Church Street and Five World Trade Center. I stood about 40 feet from what I guess you could call the ground, only it wasn't the ground in the traditional sense.

First thing I did, when I burst on to that open-walled balcony, was get back on the radio to Mark Ferran. 'I'm out!' I shouted into my handi-talkie. 'I'm out! I'm right here!'

Of course, poor Mark still had no idea where *right here* was at that point, so I sounded the bullhorn again, this time into the open air. And this time, he could hear it! He still didn't know where the hell I was, or how to reach me, but at least he could hear the siren and point his guys in our direction. I sounded it again, and again, just to be sure. Plus, our radio transmissions were now super clear. Before, all along, they'd been scratchy and far-off, but he was coming in loud and clear, and I took this as a great, good thing. At this point, it was actually Glen Rowan and his guys, Jerry Sunden, Jim Lanza and Tom Corrigan, who were closest to where we stood, so Mark communicated our location to them and worked to get them closer.

Right on my heels, once I reached the top of those stairs, were Mike Meldrum, Sal D'Agastino, Tom Falco and Matt Komorowski. Jay Jonas and Bill Butler stayed behind for a bit with Josephine Harris, because there was no way she could make it up to that landing, and no way this company was leaving her behind; soon, David Lim and Mickey Kross climbed up to join that group, and once they did, Jay Jonas and Bill Butler pushed ahead to our landing. Lieutenant Jim McGlynn hung back on the second-floor landing with his proby, Rob Bacon, trying to figure a means of escape for Coniglio and Efthimiaddes. We'd have to figure the same for Josephine Harris, some other way to lift her out of there, but these other guys from 6 Truck had no problem; they saw the path I'd taken and followed it best they could. I held out my hand to help them up the last few steps, and as they reached, one by one, they took in the scene, same as I'd done. Wide-eyed, incredulous, blown away.

It was at this point, as these guys from 6 Truck were registering what had happened, and where we stood, that Jay Jonas leaned in to me and said, 'You know, Billy here, he's got a phone.'

He gestured towards Bill Butler, like he was asking me if I wanted to bum a smoke or something, like it was the most natural thing in the world, to place a call at just that moment. Usually, I grab a phone and take it with me to each job. Like most chiefs, I leave one in a charger in the rig for just this purpose, but for some reason I forgot to take it this morning, and once forgotten it was out of my mind. It never even occurred to me to ask if any of these other guys had a phone, while we were trapped in that dark stair-well, while we were inventorying our tools and our gear. Furthest thing from my mind. If I thought about it at all, I guess I figured if someone had a phone he'd have offered it up. I don't know that the thing would have worked, trapped beneath all that mess, with most all cellular service cut, but it would have been good to know we had one, right? That's why we had our roll-call and our gear check, to get a fix on our resources.

The cells were down, all over New York City. You couldn't get a number in the 212 area code, or 718, or 914, but for some reason you could dial out to the 845 area code, which was up in Orange County, where I lived. A lot of these guys lived up there, Billy among them, which was why his phone was still working; it was programmed for a cell some 75 miles north. These guys put in calls to their wives, told them they were okay, told them what had happened. Billy even asked his wife to get on to dispatch, give them a heads-up on our whereabouts. It was real emotional, but underneath all that emotion I was quietly pissed no one had told me we had a cell phone, during that whole time. I couldn't see my way past the frustration to the human side of these exchanges. I could be a real hard-ass at times, and I guess this was one of those times. I mean, how the hell could this have been an afterthought?

When they were all through, Bill Butler offered me

the phone to call Debbie, but I couldn't bring myself to call her. What would I say? 'Hi, honey. I'll be a little late getting home tonight. Don't hold dinner.' I dialled a couple of fire department numbers instead. It seemed like a more productive use of the technology. I tried the dispatcher, the battalion, the division, but there was no getting through. I even thought to call one of the volunteer firehouses up in Orange County, thinking maybe they could get word to someone in the department, let them know a location on us, get another line of communication going.

As far as I was concerned, we were still trapped. I was hopeful, things were looking a whole lot better now than they were just a few minutes earlier, but we were a long way from safe and sound. Five World Trade Center was fully involved, Six World Trade Center was roaring pretty good, and behind them Seven World Trade Center was teetering on collapse; it seemed impossible to chart a path across these mountains of twisted steel and shattered concrete, especially with these compromised high-rises running interference. Plus, in all likelihood, Debbie didn't even know I was missing, so what was the point of me worrying her now, when there was still other stuff to worry about? We'd never talked about what to do at a time like this, whether or not she'd like to be called, if there was an opportunity to call, but I kinda thought it'd be easier on her if I kept her in the dark, for this next little while. Tell her the whole story later, when I was safe.

For the most part, moments such as these, our wives knew the drill. Too damn well, they knew the drill. If anything awful happened, someone in the department would get a chief to go out to the house or to their place of work, pick up their priest or rabbi or minister along the way. That's always been the recurring nightmare of a firefighter's wife, to get that knock on the door and see a chief and a priest on the other side. After 11 September, that nightmare has been replaced by another. There weren't enough chiefs and priests to go around, the department was so hard hit, and the new nightmare became never knowing what had happened to your husband, never getting that difficult, but at some point welcome, visit from the chief and the priest. And what kills me now, as I set this to paper, three months after the attack on the World Trade Center, is that there are still wives out there waiting on some sort of official notification from the department. A visit. A telephone call. A letter. Something. These strong, hopeful women keep calling their husbands' firehouses, and they're awash in unofficial word, but nobody has any record of where anybody was in those buildings. They're waiting, and waiting, and no one is coming. That command board I wrote about earlier, the one that had been set up in the lobby of the north tower, that's lost in a sea of rubble, so the administration is just flying blind, trying to retrace all those steps. And these poor women are just beside themselves, trying to piece together their husbands' last moments, desperate for some kind of word, some closure. And the knock on the door never comes.

I couldn't see putting Debbie through even the smallest piece of uncertainty, and on the likely chance that she was not yet worried about me, I put off calling her. As it turned out, she didn't have any idea that I'd gone missing at this point. Stephen, Lisa...

nobody knew. Even my folks out on Staten Island. Nobody knew. They all had some idea, they all let their minds race to a million 'what if?' kinda situations, but no one knew a thing, and it was easiest just to assume that I was all right. Debbie was working at the hospital, one ear tuned to what was happening at the World Trade Center, but she was doing what she could to keep her mind off her worst fears. What eventually brought her down to reality was a call from Judy Jonas, Jay's wife, to Stephen at home. Stephen then called Debbie at the newborn nursery and relayed to her the news that Jay and I had been trapped in a stairwell for several hours but that we were apparently okay. Debbie then called Judy.

Stephen had driven himself to school that morning, and when school let out early because of all the uncertainty downtown he simply headed home. He'd seen the visuals on the television, heard the reports on the radio, but like his mother he went into no-news-is-good-news mode. If he thought about me at all, during that morning and afternoon, he supposed I was at the firehouse, several miles from the scene. He knew enough about the department, and the way companies are dispatched to big jobs, that under normal circumstances I wouldn't have been anywhere near the World Trade Center. But these weren't normal circumstances.

It was my daughter Lisa, actually, who had the most calamitous morning in our immediate family, after mine, and it definitely rates some detail here. She was looking directly at the south tower when that second plane hit, at just after nine o'clock in the morning, from just outside her Pace University dormitory at the corner of William and Fulton. She somehow moved to a spot on Park Row, right across the street from City Hall, and she held that spot until the second tower came down. She saw everything. As difficult as it was to watch, she said she was powerless to look away. She even noticed the firemen racing over the Brooklyn Bridge into Manhattan, as they craned their necks out of the truck windows to glimpse the atrocity that lay in wait, and found time to wonder if there was anyone watching her dad race downtown to the biggest disaster of her lifetime. She thought of all the brave men who would die that day, and prayed that I wouldn't be one of them.

From Park Row, she told me later, everything was a blur, and underneath that blur was the certainty that I was somewhere on the scene. She hoped I'd be somewhere safe, but she couldn't know, and she couldn't bring herself to find out. At some point, early on, officials evacuated all of Lower Manhattan, but before leaving the area Lisa ducked back into her dormitory to retrieve some t-shirts, which she wet and shredded and started handing out to all the kids, so that they could breathe into them and move about without choking. The dust cloud was enveloping the entire area, and she was a true fireman's daughter, going out of her way to help those around her – and knowing just what to do. She even popped into a girlfriend's room after noticing a working computer and sent out an e-mail to all her friends, asking whoever caught the message to call her mother and tell her she was okay. She'd been trying to call herself, but couldn't get a line out of the city, and she figured Debbie had enough to do, *not* worrying about me, without also *not* worrying about her.

When Stephen got home from school, he found

about 20 messages on our answering machine – and almost all of them were from Lisa's friends. Her boyfriend. A high school friend. Another friend. And on and on. Then her e-mail got through. Stephen called his mother at the hospital and told her Lisa was safe. Debbie cried a huge sigh of relief. She'd been worried about both of us – Lisa, front and centre, and me somewhere in the back of her mind in the places where she thought the things she didn't like to think about. Once she knew Lisa was okay, that left only those unacknowledged worries about me, and Debbie was able to push these from her thinking without too much trouble; after all, she'd had a lot of practice, telling herself I was in no real danger, or that I was too smart, or too careful, or too good at my job to get hurt.

Eventually, Lisa did manage to raise Debbie on the phone.

'Don't worry about Daddy, he's okay.' Debbie kept insisting, not fully believing it herself.

'Is he working?' Lisa wanted to know.

'Yes, he's working,' Debbie said, 'but he's way up town. He's nowhere near there.'

'What are you, nuts?' Lisa shot back. 'Of course he's there. Where else would he be?'

They went back and forth like this for a while, Debbie doing what she could to calm Lisa down, Lisa doing what she could to get Debbie all frantic, and by the end of the conversation both of them were whipped into a pretty good panic.

I didn't need to know any of this at just this moment, and I didn't need either of them to know I was still nowhere near home free, so when I was finished trying to get through to the department, I handed the phone back to Billy Butler and considered our next move. At about this same time, Glen Rowan and the rest of his company were inching closer. Remember, Mark and Glen had actually splintered off into two groups some time earlier, thinking they could move more swiftly in smaller numbers, and it was Glen's group drawing nearer. A third group broke off from the original nine-man team to assist other individuals on the streets surrounding the plaza, but these two groups continued their approach. Mark met up with a firefighter from Rescue Co 2, and he had another rescue worker in tow, trying to find a path through the rubble field, so he spent some time helping these guys retrace his steps. Glen and his men could see the rear wall of the B stairwell, the side that had not been exposed, and all that was left was for them to navigate their way around. I kept blowing the siren, and I kept getting back on the radio that they could hear it, but we still weren't seeing each other.

We didn't know it just then, but these guys had been dropping their tools on the way in. It was such a gruelling climb and descent, the extra weight started to bog them down, so they dropped some of their excavation tools, their masks and cylinders, whatever they felt they wouldn't need. The air was thick with smoke still, but it wasn't the kind of heavy, acrid smoke that made breathing difficult, so they figured they could do without the masks. They were used to breathing that shit. Plus, they needed their hands to climb.

We were getting progress reports on the radio every few minutes now, and it was clear they'd be upon us in just a matter of moments. But I started to think we couldn't wait on them any more. It wasn't necessarily a logical thought; we didn't appear to be in any immediate or pressing danger, the three walls still surrounding our stairwell seemed relatively intact. In fact, we were probably safer there, protected from the roaring blaze of the surrounding buildings, than we would have been if we had tried to make our own exit, but I was overcome with this restless, antsy feeling. I wanted out, and down, and soon, and I didn't care how close Mark Ferran and Glen Rowan and the rest of the guys from 43 Truck were to the rest of us.

We'd come this far on our own. We'd make it the rest of the way, too.

ELEVEN: Home

For well over an hour after reaching the open air above our stairwell shell, our impulse had been to stay put, to wait on this last push orchestrated by Mark Ferran, Glen Rowan and these hard-charging guys from 43 Truck, but in the waiting it occurred to me that these firefighters were no better equipped to guide us down and through this treacherous terrain than we were ourselves. It occurred to all of us on that blown-out balcony, I think, and gradually we all shifted our thinking to a more active course.

Understand, we'd been prepared to be passive, to wait for an extraction team of some kind to help us navigate our way out of there, and to pull out Josephine Harris and those two guys from 39, but we were all beginning to realize that passive would kill us. Me personally, I got there by degrees. The longer I stood looking out over the north-east corner of what had only hours before been the World Trade Center complex, across a desolate sea of wreckage and ruin, fires raging on all sides, explosions that we later discovered were small arms ammunitions detonating from the Secret Service bunker that had been housed in Five World Trade Center, the more it troubled me that we were still waiting. Waiting on what, I became less and less sure. We'd get progress reports from Mark Ferran on the radio, or from Glen Rowan directly, only at some point they started to seem more like plain old reports than progress reports. Nothing against the extra efforts and back-breaking good turns of these good people – because, truly, they were all we had to pin our hopes on – but it started to seem that there was no getting to us, at just that moment. Every path they chose appeared blocked, every way in was a dead end, and in the endless re-routing we were stuck, like flies on flypaper; daylight or not, we were still trapped.

What struck me most about this scene as I looked down and out across the rubble was that there were no people. No dead, no wounded. No firefighters or other rescue workers. Nobody. It was like a shocking still-life painting. No sounds coming through over the radio, other than the voices of Mark Ferran and Glen Rowan. And no ambient noise, either, beyond the hiss and roar and crackle of the nearby fires, and the constant explosions from the grenades and other ammunition being set off in the Secret Service bunker. No lights or sirens or any other signs of rescue activity. We stood there, and we stood there, and we stood there, and at one point I thought to myself, 'Where the fuck's the cavalry?' Really, we'd been through hell, we'd hauled ass up to this pinnacle, this place of possibility, and it started to seem as if those guys from 43 might never be able to get from where they were to where we were – not anytime soon.

And so I called down for the line of roof rope we had among our gear, on the landing below. The first group of us to reach to the blown-out balcony opening hadn't thought to bring the rope along, but one of us remembered we had one on hand and it found its way up to us. I thought the rope would be key. One hundred and fifty feet long, tight nylon, tested up to 6,000 pounds. Essential gear, for any rescue or evacuation team, and essential here, for making the uncertain climb down on to this rubble field. When used alongside a life belt, it gave us enormous flexibility and freedom of movement, getting in and out of difficult spots. It used to be we had an even greater measure of freedom, when we wore a harness as part of our bunker gear to assist in our climbing, but the harnesses had a rated life expectancy of ten years and at the end of those ten years our commissioner determined it would have cost too much money to replace them. The determination was based on the fact that the harnesses were rarely used, which was true. I can probably count the number of times I had to use mine in that ten-year stretch, or assisted another firefighter in the use of his. But even though they were rarely used, they were always worn. They were a lot like those sling-seat harnesses you step into when rock-climbing, and they were an integral part of our bunker gear; the mere fact that we had them enabled us to attempt certain manoeuvres we might never have considered.

It would have cost a bundle, at about 50 or 60 dollars each, to replace these harnesses for some 8,000 firefighters. About half a million bucks. But, to my thinking, it cost us far more in missed opportunities, to jettison a piece of valued equipment. And it wasn't just that the department wouldn't order new harnesses to replace the old ones. They 'retired' the old harnesses, and had us use these old-fashioned life belts instead. The same belts we were using when I first signed on as a proby. The same belts we replaced ten years earlier, when we sprang for the harnesses.

The life belts were like thick leather weight-lifting

belts, worn tight around the waist like a girdle, with a giant hook that usually aligned with the belly or the small of the back. As a piece of climbing apparatus, it was from somewhere near the Dark Ages, nowhere near as useful as the more modern, more effective harness. We didn't wear these belts on our runs, the way we did the harnesses, but we carried them with the rest of our gear, usually a couple per company, and it really was a throwback to have to use these things, after operating so long without them. Plus, it was a big-time nuisance, to have to tote them along, instead of just stepping into the harness and forgetting about it, but they were all we had so we used them. And now we had one on that balcony, looking down on to the rubble field. We had the roof rope, packed tight into its sack for rapid deployment. The way we packed those things, with special knots tied on each end, it fed out quick and sure. We used a knot called 'a bolland on a bight' – one knot on top of another to make a third knot, any firefighter could show it to you. It's cinched in such a way that you can slip your legs into the formed loops; a half-hitch and a slippery-hitch around the chest offer further security. You're not going anywhere in one of these; it's a good knot, the same knot and harness Captain Paddy Brown, the most decorated firefighter in recent New York City history, had his team use directing the legendary 'Times Square Rescue' of 1991 when the then Lieutenant Pat Brown's crew lowered two separate firefighters to save people trapped by fire. Sadly, among the 343 brave men who died on 11 September you will find the name Captain Patrick Brown.

But we only had that one rope, which meant we'd be getting down from here one at a time – and 150 feet at a time, at least to start. The key question was which one of us would lead the way. It would likely be an arduous climb, toughest on the guy in the lead, so I wanted to give the rope to the person who'd put all of us in the best position to get out of there. Naturally, I wanted to take the lead myself, but I only wanted the lead if it made sense. I looked over at Jay Jonas and his men from 6 Truck. Jay was about six foot two inches tall, maybe 250 pounds – and he was one of the smaller guys in his company. I had nothing on any of these firefighters in terms of brute strength or sheer size, but I was agile, and I was in good shape.

Some folks have since asked me why there wasn't some type of air rescue. It was never discussed, never even considered. There were raging fires all around. A helicopter would have kicked everything up. The wind, the fine all-over powder, the smoke. The whirr of those chopped blades would have intensified the fire, and put everyone at risk. It wasn't feasible. The abutting buildings were fully involved, the exposures were vulnerable. It would have been one hellish hover, to bring a chopper in there, low enough and long enough to lift us out.

The only way out would be on foot, so I made a decision. I would take the lead on the climb down. What the situation called for was endurance, not brute force, and my objective assessment told me I was in the best shape of our bunch, best equipped for a treacherous, tentative trek across an uncertain rubble field. Of course, my objective assessment wasn't worth much, at just that time, because I wanted that rope around my waist no matter what, but looking at the situation after the fact I am convinced I made the right call.

Jay Jonas wasn't too crazy about the idea. He didn't mind that it would fall to me, the climbing down; what he minded was the climbing. 'You're nuts, Richie,' he said, as I shook out my turnout coat and strapped on the life belt. 'One slip and you're dead.'

I looked around, and wondered if maybe Jay was right. For the time being anyway, we were protected from the fires and the smoke, and the small arms, and the threat of secondary collapse from the adjacent buildings. So I agreed to wait a couple of minutes more. And those couple of minutes stretched into 10, 15, 20...

I got back on with Mark Ferran, told him I wasn't shoving off just yet, and he reassured us that Glen Rowan and his few guys were just beyond our view. 'They're coming, Richie,' he kept saying. 'They're coming.'

But they weren't coming. No one was coming. For the longest time, no one was coming. They were trying, pushing towards us from every possible angle, but they weren't getting any closer. It must have been going on two hours, and there wasn't the slightest flicker of activity. It was like hunting, when you fix your eyes on a clearing, and all you're doing is looking for movement. You focus in on a picture, and the only thing that registers is a change in that picture. That's how it was with me on that balcony, looking out across the north-east quadrant of that rubble field, anxious for some kind of blip or twitch or change to upset the scene. But the only movement across that field was the thick black smoke that covered it, as it danced and shifted with the wind, and it got to where our patience was about running out. We'd been prepared to wait, but we couldn't see what we were waiting on.

Just then, just as I tightened the life belt and worked the roof rope through the eye-bolt on the leather, I caught a trace of the movement my eyes had been seeking. There, over a rise of debris, I could make out a helmet, bobbing up and down through the field. Coming closer. Yes, definitely, that's what it was. A helmet, coming our way – caught behind a billowing of smoke, but here and there visible through the clearing. There was no indication that the guy wearing the helmet had spotted us just yet, so I lifted the bullhorn to my lips and called out to him. 'Over here!' I said. 'Over here!' Then I sounded the siren for good measure, and with this the guy looked up and noticed. He waved his arms, to indicate we'd been spotted – and then, from Mark Ferran on the radio, we heard that they were just a few minutes away. A few minutes, a few hundred feet. Getting close. And as they approached, my heart soared and then sank, in pretty much the same motion, because as these guys from 43 came into view I could see they were empty-handed. No tools. No rescue equipment. Nothing. How the hell were they gonna get us out of here? With what?

What had happened, I later learned, was that these guys had been working their butts off for going on two hours, crawling in and out of burning buildings, climbing up and down these massive craters of rubble and debris. They were literally exhausted, and every here and there they'd drop a piece of essential life-saving equipment to lighten their load. It was the kind of climbing they needed their hands for, to grab on to this beam or that pipe, or balance against that sidewall. Anyway, it was all these guys could do to lift their own feet at this point, forget lugging 50, 60, 100 pounds of equipment. They'd started out ferrying all kinds of additional weight, and at some point their goal became simply to reach us, and to worry about getting us out of there later.

Finally, Jerry Suden, Ladder Co 43, was in close enough to where he could hear me through my bullhorn. He stood across from us, in the middle of a mini-crater; he'd climbed down 50 feet or so, and was now looking at climbing back up towards us, another 50 feet or so. The space between us could have been covered with a couple of flaps of a pigeon's wings, but on foot it could have taken a half-hour. Still, Jerry didn't like the look of the terrain going forward, didn't like his present position, and we couldn't blame him. From where we were, looking down from our perch several storeys above that particular piece of rubble field, we could see he was in a tough spot. The buildings just behind him and to his left were looking like they too might collapse at any time, and there were whole chunks of concrete falling to both sides. Flames dancing everywhere. The small arms detonations were kicking up a notch or two, and it sounded like this poor guy was being fired at, by snipers or some unseen terrorists, at close range. It must have seemed to these guys like they were crossing a minefield.

And so we made to move. The men of Ladder Co 43 would reach us eventually, but once they did it became a game of tag. Us and them. And they were now 'it', meaning it now became their turn to sit with Josephine Harris, to take responsibility for getting her out of there, to figure a way to get to Coniglio and Efthimiaddes in that lower void. To dig down to Chief Prunty, determine if he was still alive, and find a way to get him out too. These poor guys were thrashed, and could use the moment to catch their breaths and their wits. The group of us on that balcony perch were ready to move – myself, Mike Meldrum, Sal D'Agastino, Tom Falco and Matt Komorowski – and for the moment, as our rescuers stood across that small crater and wondered how to reach those final few feet to where we stood, I determined that it was easier for us to get down from where we were than it was for them to get to us. I was already tied on. Visibility had improved dramatically since the collapse of the north tower, but it was still spotty. Every here and there we'd get these waves of dense, grey-black smoke, hugging close to the debris field, but it moved with the wind and opened up soon enough. It was kinda like being up in the clouds, on an otherwise clear day; you pass through the clouds and into blue sky, only here the clouds were made of thick, killing smoke.

Right away, I started searching out a path to take me down from that balcony, the way you'd look for a line down a mogul field when you're skiing. You start thinking two, three, four turns ahead. Shimmying down this pipe here, tip-toeing across that beam there, pulling up by that jagged edge to get a toe-hold on that girder – taking the bumps one at a time but always thinking about the next one, and the one after that.

During that whole time we'd been waiting, at the back of my mind, I was mapping out an exit route, and here I was, putting it into play. I moved slowly, but I moved. It was like climbing down from the top branches of a tree, always looking for a good place to rest or pivot. I ripped my pants on a piece of jagged metal almost as soon as I started out, but for the most part I made decent progress – and precious few mistakes. This was the kind of situation where one misstep could have put me in serious jeopardy, so I didn't plan on making any.

When I reached a point about ten or 15 feet away from the balcony, I tied off the first section of rope to what had once been a post of a stairwell. Actually, I only assumed it was a post, it could have been anything. That was one of the strange, unsettling things about the climb out of that rubble field, seeing all these objects and scraps and trying to figure out what purpose they had served.

Next, I took the remaining 140 feet of rope and doubled it up on itself and started looking for another substantial object to tie off against. In this way, I meant to lay a fixed line for everyone else in that stairwell to follow.

We knew it'd be tough going, but we had no idea how tough. The fires, the precarious footing, the intense, all-over fatigue. The bombs bursting in the air – meaning those weirdly disturbing small arms detonations. And the smoke! I never could determine what it was about that smoke we were breathing on the way down but, as I've written, I had a pretty good nose for virtually every type of smoke. Here, though, there was something especially irritating and sharp in the air we were made to breathe, and I couldn't place it. For days afterwards, I couldn't think what it was, until someone suggested there might have been different types of tear-gas bombs going off in that Secret Service bunker. That would have certainly accounted for the difficulty I has having drawing a full breath, and the pain I was feeling in my eyes. There were times it seemed as if I'd be blinded by all that smoke, by the burns and abrasions to my eyes, but I managed the pain to where I could still see, in fits and starts.

Our balcony perch hung about 50 feet above the main surface area of the rubble field, which meant that my first advance was mostly down, almost like I was rappelling down a sheer, craggy face of a mountain, except I never really had to leave my feet and sway or anything like that. But there was a steep pitch to that initial descent, and the trick was leaning into the fall and letting the rope do the work and cutting the easiest path from one point to the next. I don't know that I succeeded completely but I succeeded sufficiently. I reached to a solidly entrenched beam, and found a way to tie off another section of rope, and then I reached a girder that had somehow impaled itself into a pile of debris and tied off another section there, and in this way I managed to lay a fixed line for my brothers from 6 Truck to follow. They wouldn't be wearing the life belt as they made their ways down, so this fixed line would be their only tether. Occasionally, I'd have to catwalk across a horizontal I-beam, like a tightrope walker, and the thing would be coated with six or eight inches of cement dust and ash, and as I crossed I worked to clear the beam of that fine powder with each step, almost like I was shuffling. My goal was not only to move down from that perch on to more certain footing, but to cut the easiest possible trail for the others to follow.

Eventually, and soon enough, I ran out of rope, and at this point the others started to follow. One by one, with Mike Meldrum next in line and Sal D'Agastino after him, and then Tom Falco and Matt Komorowski. They'd been watching – all of them, all along – but no one was saying anything. No one was eye-balling the field from on high, suggesting one line of descent or another, even with their better vantage point. They just watched, and quietly rooted me on, because the hard truth of it was that if I didn't make it down, neither would they. At least, not with this rope, and it was the only rope we had.

Once these first four guys from 6 Truck reached the end of the rope, the initial rescue party from 43 Truck began moving back up those same fixed lines to the open stairwell. They relieved Jay Jonas and Billy Butler, and continued looking after Josephine Harris until additional help could arrive. This shift was important, because Jay and his men had made a pledge, as a company, to stay with this woman until the end, and here it was pretty damn close to the end. Close enough, anyway, for 43 to take over.

The plan was for Meldrum, D'Agastino, Falco and Komorowski to continue their descent while I waited on Jonas and Butler at the bottom of the fixed line, but as they set off we could tell there was no certain path for them to follow. They'd have to move by trial and error – and as it happened their first few steps were made mostly in error. They moved at first towards the raging fires of Five World Trade Center to our east, but as they did Mark Ferran came back on my radio cautioning us from that building. 'You gotta get out of here!' he shouted. 'We're losing the building! We're losing the way out!' That had been their way in, through that Borders bookstore, so naturally it had loomed as our way out. To the north and north-west, there was the teetering inferno of Six World Trade Center; the crew from 43 Truck had been that way before and had been turned back. To the south, there was a vast, treacherous, endless rubble field, and the perilous high-rises of Two and Four World Trade Center, each one a secondary collapse waiting to happen. To the west and south-west, there was rubble as far as we could see – and beyond that there was more rubble, but it seemed to be in manageable, scalable piles. It seemed our best course, our only course.

And yet after reviewing our options from his vantage point, Mark Ferran hollered, 'You can't go west, Rich. It's too dangerous.'

'You're killing me,' I said back into the handi-talkie at my chest. 'I can't go west? All I got is west.'

And so, we simply moved. In no apparent direction at first, but we couldn't stand still. Soon enough, Meldrum, D'Agastino, Falco and Komorowski doubled back to join me and the rest of their company at the base of the fixed line, and from there we set off again, as a unit, with me returned to the lead. It was a case of the blind leading the blind, because my vision was all messed up from the smoke – but even if I could have seen I wouldn't have been able to tell where I was going. I was just going.

The hairiest part of the journey came another 20 feet or so from the rubble field: the rope ran out when I reached a fallen I-beam that offered the only path over a stretch of chewed metal. The beam had fallen in such a way that it made a kind of bridge, reaching out over a small crater, about eight or ten feet above the pile. It ran into a chunk of what looked to be waste pipe, which presented the next leg of my journey, but there was a gap of about 18 inches, across and down, separating the beam from the pipe. On firm ground, such a gap would have been a short hop or skip, taken without a thought, but here on this climb down a pile of debris it was pretty perilous. It was like playing hopscotch on a staircase, taking three or four steps at a time and meaning to strike a narrow target on the next step. The beam and pipe rested unsteadily, and without a rope or another beam to grab on to I was unsteady as well, and it had been my experience that unsteady on top of unsteady was not a good fit. Still, it was the only good way down, so I went for it. I wasn't so high up that a fall from this height would have killed me, but it would have broken a couple of bones, easy. I suppose I could have hit my head. And who knew what the hell I would have fallen on. A jagged piece of something could have pierced right through me like nothing at all, so falling wasn't an option.

Somehow, I made it across – teetering the whole way, but I made it across. I think I might have willed myself not to think about it, because the truth was it was hardly a leap at all. It was a leap of faith, more than anything else, and once I had crossed to the waste pipe I wondered what the big deal had been.

But it had been a big deal, that was the point, and soon enough it was a big deal for the next guy in line, Mike Meldrum, who at six foot five, 260 pounds wasn't exactly light on his feet. Like I said, some of Jay's guys tilted the scales a bit, and if Mike was freaked by such a delicate move then surely the others would have their own troubles as well. Mike reached to the edge of the beam and just about froze. 'I can't do it, Chief,' he said. There was no panic in his voice, just the certainty that he had reached an impasse.

'Sure you can,' I coaxed. 'You do it all the time. Just think of it like a staircase with a couple of steps missing.'

He thought about this for a beat or two, looked around to catch his bearings, but he still didn't like his chances. 'No, Chief, I can't,' he said. 'I think I'll go back. Find another way down.'

'There is no other way down, Mike,' I said, 'and this is nothing. Just put it out of your mind.' It wouldn't do to have Mike thinking he couldn't take this next step, because then maybe Sal wouldn't be able to manage it, and then all the rest of them.

But Mike couldn't move. He was paralyzed – not by fear, but by doubt. He must have caught an image of himself spinning on that waste pipe, like a lumberjack in a log-rolling contest, and spilling on to that hazardous debris pile ten feet below. Mike Meldrum is a professional, he'd been on the job 20 years, but even professionals have their doubts, especially after the day we'd been having. So I reached into my bag of leadership tricks for a tack that might take Mike over this hump. 'I'll bet you a hundred dollars you can do it with your eyes closed,' I said, thinking this would spur him on. There's nothing like a challenge to get a guy going – especially a challenge backed by a hundred dollars.

Mike looked at me, and he looked at the gap between the beam and the pipe, and then he looked back at me.

'Just do it,' I tried again. 'A hundred bucks.'

He looked at me long and hard. 'You're fucking crazy!' he said.

'All right,' I said, 'then do it with your goddamn eyes open.'

'Damn straight I'll do it with my eyes open,' he said, and he took the step like it was nothing at all.

When he reached next to me, on what passed for firm ground on that piece of rubble field, Mike shook the shakes from his bones and managed a small smile. 'You owe me a hundred bucks,' he said.

'Like hell,' I shot back. 'I offered a hundred bucks with your eyes closed. You opened your eyes.'

He didn't argue. He'd got past that terrifying hurdle, and I guess he figured that alone was worth a hundred bucks.

Initially, I thought we were moving north and east, hoping to retrace the path taken by Glen Rowan and his men, at least in terms of general direction. The plan was somehow to reach back to the obscured concourse level entrance of Five World Trade Center, but it was impossible to figure which way we were going. There were no fixed markers or landmarks to serve as guideposts, and we were fat in the middle of a densely packed rubble field. When we were down at the bottom of any of the dozens of big-ass craters that had formed, we couldn't see anything above the rise on all sides. There was the bright sun, which when the smoke occasionally cleared offered some sense of direction, but it hurt my eyes too much to look skyward. The best I could manage was dead ahead, and even here I was operating with a limited view. I couldn't keep my eyes open for more than a few seconds at a time, without having to close them against the agonizing abrasions and irritations; the light seemed to make things worse. Out in front, then, and at my feet, there were nothing but great hills and steep valleys of debris, and up above there was one burning building after the next. And, in the burning, these buildings looked nothing like they had in memory, so for a while in there we were just moving. Our only marker, really, was the shell of stairwell we had just left, as it rose about ten storeys from the ashes of the north tower – and my one working sense of direction was down.

Wherever we were, however we got there, we were in an untenable position. The smoke was becoming too caustic. We couldn't breathe. We couldn't see. I don't know how it was for anyone else who was descending in my wake, but my eyes were burning something fierce, almost to the point where I couldn't see a blessed thing. And we stood dangerously close to what on any other day would have been one of the worst high-rise fires in New York City history. Actually, it would have been *two* of the worst fires in history. Five and Six World Trade Center were chillingly ablaze, like something from a computer-animated special effects shop in Hollywood. I'd never seen anything like it, the way those buildings were roaring. There was no fighting those fires. There were no men to deploy, given the devastation all around, and no good means available to fight those blazes, but we'd have given up those buildings even if we were operating at full strength and in ideal conditions. At least that's what I would have done, if I'd been in charge. Some fires you fight, and some you contain; you worry about protecting the exposures, get the office workers and rubberneckers out of there, let the fire have its way.

On the radio, from his ever-changing posts in and around the less compromised perimeter buildings,

Mark Ferran continued to caution us from heading west. That's the direction he'd first tried to come in from, and there had been no clear path. He'd doubled back more times than he could count, and the buildings behind him were now in a more compromised state than they had been on the way in. Mark had spent the past half-hour frantically searching for another way to reach us, which would then serve as our way out, but he couldn't catch a break. Even from our perspective, we could tell there was no obvious way out of that rubble field. East, we could see, was out of the question, because of the way the L-shaped Five World Trade Center stood on the corner of Church and Vesey, an inferno waiting to fall and take us out with it. South appeared to be mountain after mountain of packed debris, interrupted by crater after crater; for every piece of progress headed down, there'd be another piece of regress headed back up. There'd be no climbing out in that direction, not by nightfall anyway.

A note on the time. Best guess, at this point, we were looking at three o'clock in the afternoon. Approximately six hours after the first plane hit the north tower that morning. The sun was still fairly high in the bright September sky, and every now and then, when the smoke shifted and created a pocket of brief visibility, I could see all the way to the heavens and marvel at the clear, clear sky. That is, I could catch a glimpse before my eyes hurt so badly I'd have to close them and imagine the scene. My God, what a stellar day it had been, before everything turned to shit. And here we were, teetering on our last legs, bent and broken same as the compromised buildings that contained us, wondering how the hell we would ever get out of there. And when. The truth was, there wasn't much sense of direction left in any of us, and I wound up going pretty much directly against Mark Ferran's recommendation, at least at first. I hadn't meant to, but the path I wound up following, immediately after descending from that balcony, was east and then north and then west, so that I wound up circumnavigating that stairwell shell and blazing a path that seemed to want to take me south and west. It's like I was being pulled along, without much room for a plan. With each step, I merely considered the easiest push forward, with hardly a logical thought for direction, and in this way managed to put some serious distance between myself and what was left of One World Trade Center. The other explanation for all my wandering and misdirection was that I was having all kinds of trouble keeping my eyes open, and the irony of that situation wasn't lost on me: there I was, leading our escape, and I couldn't see for shit.

As I moved further and further from that still-standing stairwell, which was now filling with those other guys from 43 who climbed our left-behind length of rope to reach to Josephine Harris, Jeff Coniglio, and Jim Efthimiaddes, I allowed myself to get lost in thought, a little bit. Well, maybe 'lost in thought' is putting an over-emphasis on things; it was impossible to develop a rhythm to our climb, as each tentative step required a plan, but it was possible to let all these path-charting thoughts share space with other, more historical matters, to put these motions in context. I thought about where I stood, on this bizarre, giant, unprecedented field of wreckage. It threw me, to think about where I was, what had happened, what I was doing, trespassing on this fresh burial ground. I thought of all those lives underfoot, thousands upon thousands of them, and the ways those lives were connected to hundreds of thousands of other lives far away from this field. Yet I couldn't see a body or a fragment of clothing or anything to indicate a human presence. For miles around, I later learned, people were turning up papers and other inter-office artefacts from the upper floors of these towers, but where I stood there was nothing to suggest what had gone on in these buildings up until just a few hours earlier. There was nothing human, nothing personal. Nothing *real*. All I saw was steel and concrete rubble. Every specific thing, every identifiable thing was just disintegrated, pulverized. Gone. And we were left to walk across what was left of it, me and these six warriors from 6 Truck.

Gradually, I pointed us towards the path of least resistance, and the guys followed my line of descent. For the most part, anyway. It's not like I was leaving footprints for them to trace, so there was some veering off here and there, but we all moved in the same general direction, like a stretched-out conga line. I pushed all thoughts of mortality aside – mine, and everyone else's – and pressed forward. It was the only way to get past what we had to get past.

As we moved, and encountered one obstacle after another, a couple of Jay's guys started reaching into their pockets for the personal safety ropes they carried, and they used these to help each other over particularly shaky spots. These personal ropes were quarter-inch nylon, 25 feet long, able to support the weight of the heaviest firefighter in the department. They used to be standard equipment, part of our bunker gear, but these too were recalled after their ten-year rated life expectancy came due, and were never replaced. The brass actually made us turn them all in – every battalion had to account for every last rope – but since it's a low-cost, lightweight item that can pay big dividends, a lot of guys had since chosen to buy their own and carry them on their belts or in their pockets. Even when we dropped our gear, for greater freedom of movement, they hung on to these ropes. The things were so light and out-of-the-way, most guys probably didn't remember they were carrying them.

I didn't carry a personal rope, and I was way out in front, so I was left to negotiate these turns on my own, and I must confess now, in hindsight, I got a little reckless as I made my way out of that rubble field. Not reckless as in irresponsible or anything like that, but I wasn't moving as gingerly as these guys, that's for sure. I wasn't as tentative as I'd been when I first set off from that ripped-open balcony. It got to where I was taking steps without thinking about them, trusting my gut, trusting that the rubble beneath would support my weight. I moved on faith and instinct, whereas the rest of our group were perhaps a little more cautious, watching each other's backs, assisting each other with the ropes, worrying over each step, so that eventually I opened up a good deal of distance between myself and the rest of the pack. This was okay. In my head, it moved from being a leadership situation, where as chief it fell to me to get these guys out and down, to being a question of survival. I was being trailed by a company of men, and these guys naturally felt a certain responsibility to each other, but my focus shifted from taking charge and assuming command to getting the hell out of there. We still had people trapped up in that stairwell, and I started thinking that the best way to get them the help they'd need was to reach down to one of these perimeter streets, grab some firefighters by their collars, tell them what had happened and what was happening still, and point them in the right direction. I had to get help to these people, but I was too spent to help them myself. I had nothing left.

There was very little activity south and west of where the north tower had stood, approaching the shell of the old Vista Hotel and the South Bridge that had recently connected the World Trade Center plaza to One World Financial Center, across West Street. With the first collapse, the makeshift command post had been pushed to the north-west, at the corner of West and Vesey, which was why Mark Ferran's initial push had come from that direction. Down here, by the South Bridge, which was now covered to its base with a pile of scattershot debris, there was no real rescue presence; there were dozens of rescue workers, but there didn't seem to be any organized effort; everyone was just wandering around, waiting to be told what to do. I was screened from the department command post by several walls of ruin, stretching several city blocks. I did, however, pass two guys from Rescue Co 2, headed up a monstrous crater towards the stairwell, about 45 minutes after I'd set off from that fixed line of roof rope. I assumed they'd been circling the area, looking for a way in, and when they reached the shell of the hotel they saw their opening. They actually saw me first, these firefighters, and started waving. When we reached shouting distance one of them called up to me and said, 'Hey, buddy, can you give us a hand?'

I thought, 'What the hell is this guy talking about?' Clearly, he didn't know who I was, or what I'd been through. I no longer had my helmet or my turnout coat, so from a distance there was no way to know I was a chief, but more than that he couldn't have guessed at my situation. There wasn't enough back-and-forth on the radio to suggest that our rescue and evacuation was any kind of priority. Hell, it was entirely possible these two guys didn't even know there was a specific effort underway. Given the suspect nature of our command and communication structure, they probably didn't realize that there were people trapped – other firefighters – but it struck me as absurdly funny, at just that moment. To have stood between the seventh and sixth floors of this giant skyscraper as it collapsed around me, to have survived black, terrifying hours in an uncertain tomb, to have clawed my way up 50 feet of jagged metal and busted concrete to reach open air, to have ignored the advice of other firefighters and climbed down this mountain of debris... all to bump into these two guys, just starting out on fresh legs, looking for an assist.

As I reached closer, I filled these guys in on what I knew, pointed them in the direction of the stairwell, showed them the path I had just taken. One of them gestured to the life belt I still wore around my waist, and asked if he could take it in with him. 'You won't be needing that, Chief,' he said, 'here on in.'

So I gave them the belt, told them where they'd find 43 Truck on the radio, told them about Coniglio and Efthimiaddes, and Chief Prunty, who I hoped was still alive. I wanted to make sure they took the equipment they'd need to help Josephine Harris out of there. Mickey Kross, Rob Bacon, David Lim and Jim McGlynn had hung back, but they could move out on their own power, if they hadn't done so already. It was these others who needed attention. And quick.

And then I continued on, pressing towards the remains of the hotel, and the swallowed-up bridge. All along, with my eyes half- or occasionally fully open to the scene, I scanned the ground at my feet, waiting for the dense rubble to give way to actual pavement. The further I moved from the north tower, the more firefighters I caught in my limited view, but here again they didn't seem to be doing anything but waiting on orders. This frustrated me at the time, and it frustrates me now in the recounting. We had people trapped up there. This was a rescue situation. I started pulling aside some of the firefighters in my path, asking them what was going on, telling them who I was and what had happened, and I kept getting these looks like they thought I was crazy. They had no idea what I was talking about – and as far as I knew, we were the only folks in need of rescuing. There was no saving these abutting buildings, and there were no other people to save, so how was it that nobody seemed to know about the dozen or so people who had been trapped in the central stairwell of the north tower?

Here I was, thinking I would stagger down from that pile of rubble to a red carpet welcome, with backslaps and hugs and all that, but it was a non-event, my stepping from that field. What was that old poster, from the Vietnam War? 'What if they gave a war and nobody came?' That's kinda like it was here. What if there'd been this impossible rescue, this impossible escape, and nobody knew about it? At this point, I had put so much space between myself and Jay Jonas and company that I could no longer see them, which meant that none of the firefighters by the South Bridge could see them either. I was moving along the perimeter of the plaza alone. I had my collared work shirt on, which had once been white but was now a sweaty, ash-soaked grey, and I guess it's possible these guys mistook me for just another firefighter having a bad day. I grabbed a staff chief and told him we had people trapped up on that rubble field, but he just looked at me blankly and informed me the department wasn't sending anyone in. Then I found a deputy, and heard pretty much the same. I thought, 'These people are out of their friggin' minds.' Actually, I think I even said as much, to one of the deputies. I said, 'Jesus Christ, you got guys up there! Do something!'

But there was nothing doing.

These were some of the most maddening moments of the entire day. After a few stops and starts, trying to explain what had gone on, what was going on still, I decided to bypass all this confusion and seek out Pat McNally at the command post. Pat was the deputy in charge; Mark Ferran had been in contact with him; if anyone down here was waiting on our rescue, it would be Pat and his partner at command that afternoon, Nick Visconti, who, though he had been in contact with Jay Jonas, had never got close to us in the rescue effort.

Trouble was, there was no reaching command without circling south-west around One World Financial Center, and behind the high-rise buildings on South End Avenue – a good half-mile or so out of my way, but there was no better path to West and Vesey. I couldn't tell where the rubble field ended and civilization began. I don't think I reached pavement until I stumbled on to Albany Street, headed towards Battery Park City. The surrounding streets were so severely choked with beams and pipes and concrete boulders it seemed to me it would take years to complete a clean-up, but as I moved far enough away from the points of direct impact, the rubble seemed less densely packed, until it finally gave way to the fine grey dust that coated the entire southern tip of Manhattan for the next several weeks.

As I moved about, making my way to West and Vesey, I started to run into people I knew. Eddie Meehan, a lieutenant from Ladder Co 22, one of my home companies. Oh, was he a welcome sight! And he was surprised as hell to see me, too! 'Chief Pitch!' he shouted out, soon as he caught a glimpse of me. 'Man, we all thought we'd lost you!'

My aide Gary Sheridan had been banged around pretty good in the collapse, but he was mostly shaken by the thought that I was still inside the north tower when it came down. We had a lot of our guys on the scene from our battalion – Ladder Co 22 and Ladder Co 25, and Engine Co 74 – and a group of them survived simply because they ran in the right direction when the building started to rumble. Another group didn't make it because they ran in the wrong direction. And those who'd survived were all hearing that I hadn't. The news shot through the battalion. For hours now, my guys had written me off as dead – and here I was, staggering about like a blind man, but very much alive.

At around this time, soon after I made my first

meaningful contact with someone in charge, my son Stephen heard a knock at our front door in Chester, New York, some 70 miles north of ground zero. It's ironic, don't you think, that we firefighters had such trouble communicating with each other all across the World Trade Center complex, when we were only a few hundred feet apart, and no trouble dispatching a messenger to my house to deliver the word that I was okay? I'm assuming now that the messenger was one of the 'vollies' from the local volunteer fire department, because all off-duty New York City firefighters had made their way to the scene. Stephen didn't catch the guy's name, or assignment, but it was someone from the local volunteer department, doing the good-news version of those dreaded chief-and-priest visits. It was about four o'clock in the afternoon, and Stephen had come home to our empty house a few hours earlier, after school had let out early. He called his mother at the hospital a few times. He knew his sister was okay and being evacuated, but no word from me. The phone kept ringing from family and friends who wanted to know if I was all right, but no word, so he just hoped all was right with our immediate world. The city was in ruins and chaos – Stephen had caught it all on television – but his family was just fine, as far as he was concerned, even though he had no clear idea where his father was. And then came this knock.

Stephen opened the door and was surprised to see a fireman on the other side.

'Hey,' Stephen said. 'Can I help you?'

'Is this the home of Richard Picciotto?' the fireman asked. 'Battalion commander, New York City Fire Department?'

Stephen nodded, fighting off a sick feeling in his stomach that this was not going to be something he wanted to hear.

'Is that your father?' the fireman asked.

Stephen nodded again, waiting for it.

'Good,' the fireman said, and then his demeanour brightened somewhat. 'I've just been sent to tell you your father's out.'

Stephen's first thought was, 'Out? Out of what?' He said, 'I'm not sure I know what you're talking about.'

'That's all I know,' the firefighter replied, 'to tell you he's out, he's okay. I'm sure you'll hear from him shortly.'

It wasn't exactly the most informative courtesy call in the department's history, but there it was. As soon as the guy left, Stephen was on the phone to his mother, to tell her what had happened. Like Stephen, Debbie was relieved to know that I was *out*, but she had no idea what it was I'd got into. And she wouldn't know for some time. Stephen had that brief phone call from Jay Jonas's wife and he called his mother Debbie, telling her that Jay and I had been trapped in the stairwell, but that only added to her confusion. I wasn't able to get her on the telephone until I'd been taken to the hospital, so she was still pretty much in the dark.

I was still in the dark myself, with the way my eyes were killing me. At this point, I couldn't stand to open them, so I was moving mostly on feel, and occasional peeks at the scene. First chance I got, I grabbed a bottle of water and poured it over my eyes. I don't know that it helped much, but the splash of cold offered a distraction, and momentary relief.

I couldn't admit it at the time, but I can see it now – I was in pretty bad shape. Eddie Meehan and some of the other guys in my company wanted to send me off in an ambulance right away, but I was still determined to assist in the rescue of those left behind in the stairwell. I couldn't punch out until they were taken care of. To my thinking, there was no one on the scene better equipped to lead an extraction team to the void where Josephine Harris, Jeff Coniglio and Jim Efthimiaddes lay waiting – and no one better to advocate on their behalf. It was a tough sell, to get the deputies in charge to see their way to authorizing a rescue effort. It didn't matter that to me it was a no-brainer, because these guys were making a tough decision a minute and my take didn't necessarily count. It's like I said before, how my idea of conservative clashed with their ideas of conservative. You have to consider whether you put 100 of your men on the line to save one. You have to consider that a civilian life trumps a firefighter's life, always, but that it's not always good practice to put 100 firefighters' lives at risk to save one civilian. If it were up to the 100 firefighters, they'd make that deal in a heartbeat, but the chief's job is to make sure the department still functions, to make sure the equations make sense. That was the dilemma here: those perimeter fires and secondary collapses put our people at enormous risk, but at the same time we had a precise location on three of our own in desperate need of rescue. To me, there was but one course of action: we had an obligation to move. To some, it wasn't so clear.

Finally, after a little too much back-and-forth with the chiefs assigned to the scene, we dispatched three entire companies, fully equipped with evacuation gear, to move in towards the stairwell – all led by Battalion Chief John Salka, another friend who had actually been a part of that same study group with me and Jay Jonas.

It's amazing to me now, looking over my shoulder, how many hoops I still had to jump through once I made it to firm ground. There should have been enough of a fire department presence, and a shared presence of mind, to push through these paces without any input from me. I'd been through enough. I was in tremendous pain, and completely, overwhelmingly exhausted. But the brass was dragging its feet sending in reinforcements to the scene, so I figured my treatment could wait until we had a solid rescue effort in place.

Meanwhile, my fellow stairwell 'survivors' made their way out of the rubble field – and it was as if nobody noticed. The guys from 6 Truck splintered off in all different directions. Jay Jonas actually kept moving east, and didn't stop until he reached his own firehouse in Chinatown, on Canal Street, some 20 blocks away. He told me later he was dazed, and there was no one around to help him, so he started walking back to the firehouse. He couldn't think what else to do, he said. At one point, he jumped on a city bus, to take him the rest of the way. Matt Komorowski found himself in an ambulance headed for St Vincent's Hospital, where doctors treated his separated shoulder. Billy Butler was, for some reason, ushered by emergency medical types a few blocks west to the river, and from there sent by ferry to a hospital in New Jersey. I stayed on the radio a bit, to assist the effort up to that stairwell, and in between calls I was attended to by emergency medical technicians, who did what they could to flush out my eyes and relieve the pain. They soaked a bunch of towels for me, and wrapped them around my eyes, but even with the towels I kept reaching for more water. Bottles and bottles of water, just dumping them out over the towel, and later over the bandages they gave me, just to get some relief.

Eventually, I was taken to St Vincent's Hospital, which was the closest city hospital to the World Trade Center complex. (I was loaded into the ambulance by one of my probies, Chris Dunic of Ladder Co 22, who seemed mighty glad to see me still alive and who was later sent to Afghanistan as a member of a US Special Forces unit.) The scene at St Vincent's was both creepy and heartbreaking, the way hundreds of emergency personnel had gathered to administer to victims – only there were hardly any victims. For a beat or two in there, it seemed there was just me. Matt Komorowski was there too, but I didn't see him at first. I just saw all these doctors and nurses, lined up outside, waiting to do something, and there was no one to save. There were a few people who'd been hit by incidental debris, some cuts and bruises and minor concussions, but there was no one to save. There were the 14 of us in that stairwell who survived, and not too many others.

Eventually McGlynn and Bacon assisted some of those left-behind firefighters from 43 Truck in evacuating Coniglio and Efthimiaddes – and they were mostly okay. With the right tools, which arrived soon enough, they were able to get them out of there in no time, and all four of those guys from Engine Co 39 made the trek out of that rubble field on their own power, same as the rest of us had done. Kross and Lim had been helped out of there some time earlier, and had reached safety. Josephine Harris caught a ride down in a Stokes basket sometime before six o'clock that evening – and she too was in reasonably good shape, considering.

We were all in reasonably good shape, *considering*. It's the considering that would keep us up nights. Later Chief Prunty's body was discovered.

Soon as I could, I called the house from the hospital, but there was no answer. I tried Debbie at work, but I couldn't get through. Finally I was able to reach my parents in Staten Island. I told them I was okay, that I would fill them in later, and asked them to try to get word to Debbie and the kids. There wasn't much time to talk, with the way the doctors and nurses were working me over. I was banged up pretty good. My shoulder hurt. My back ached. I was coughing like a lifelong smoker. And my eyes were shot. Cuts. Abrasions. Burns. An ophthalmologist came by and flushed them out for a good long time. It was intensely painful, and I picked up all these hushed conversations about what my vision would be like when my eyes finally healed, if they finally healed.

For the longest time, I lay there in the emergency room, my eyes wrapped like a mummy, a cold bottle of sterile water in one hand so that I could soak my eyes every now and then through the bandages. They'd cut my clothes off, as soon as I came in, put me in one of those powder-blue hospital gowns, the kind that leaves your ass hanging out the back, and that's how I was when I got word that New York City Mayor Rudy Giuliani was in the hospital, to pay a call on those firefighters who'd been trapped in that stairwell. That would be me, I figured – and Komorowski, I learned later. I thought to myself, 'Well, I don't want to see the Mayor, looking like this, with my ass hanging out and my eyes all bandaged up.' So I hung back for a bit, and waited until he'd moved on.

I couldn't hide from the fire department chaplain, though. He managed to find me, making his rounds, looking for the few other firefighters who'd been brought here. I don't know if he was working off a list, or just poking around, seeking guys to help or comfort, but somehow he found me – and I wasn't too thrilled to be found. I was all bandaged up, all but blind, lying there like I'd been through hell. And in the cool, sterile comfort of that emergency room, I started to hear the chaplain administering last rites. To me. Let me tell you, this freaked me out something fierce.

'Father,' I managed to say, 'I'm not dying.' I didn't put it in the form of a question, but with his being here and all, giving me my last rites, the question was raised.

'No,' he said, in his best comforting voice, 'but let me say it anyway.' He explained how the prayers also covered the anointing of the sick, so I figured it couldn't hurt, to hedge all my bets.

I finally got through to Debbie, and that was a real emotional call. It was probably about eight o'clock in the evening by this point, so she was all bundled up inside. Sick with worry, with not knowing. She'd heard from my folks, that I was in the hospital, nothing life-threatening, they were taking good care of me. She'd heard from Jay Jonas, who'd arrived home and was just calling to check in, so he filled Debbie in on what had happened. She'd spoken to Lisa, and gotten her perspective on the day's horrific events. Up until the time I called, she had spent the evening sitting with Stephen, absorbed by the television coverage, and each time she saw that north tower collapse, over and over, she couldn't shake the thought that I had been inside. She couldn't really get her mind around it, she told me later.

'Deb, it's me,' I said, when she picked up.

'Pitch!' she shouted. And then the tears came. 'Pitch, I can't believe it's you. I can't believe it. Are you okay?'

I told her about my eyes, told her they were messed up pretty good, and – good nurse that she is – she pumped me with all kinds of questions. I wasn't interested in all the details at just that moment, hadn't really paid attention to the doctor when he explained what was going on. All I knew was that I was in pain, and that I couldn't see, and that they'd be discharging me soon.

Anyway, it wasn't what I wanted to talk about. I hadn't realized there was something I wanted to talk about, but here it was. 'Deb,' I said, 'I'm done.'

She told me later it was the most curious thing, the way I put this one thought out there with such finality. She had no idea what I was talking about. I wasn't so sure myself. Done? With what? At first Debbie guessed I meant I was finished at the hospital, about to be sent home, but deep down she knew I meant something more.

'Done with what?' she asked.

'Done,' I said again. 'Finished. I'm not going back to work.'

Even here, I wasn't doing much to clear things up for either of us. It hadn't even occurred to me to leave the job, but once I heard Debbie's voice, the prospect just spilled out of me. Like it was the natural next move, the only next move. We'd talked about retirement, off and on, over the years, especially once I hit my 20-year anniversary in the department, which entitled me to a full pension. The guys talked about this kinda shit all the time; when they were getting out, what they would do when they got out, but it had never been a real possibility, as far as I was concerned. Not anytime soon. I loved what I was doing too much to give it up. I hated the paperwork and the bullshit and the politics of it, but I loved the job. I loved the jobs. And the guys. I was *this close* to making deputy, tenth or so on the promotion list, which would put a cap on a wild ride of a career. And there was still some firefighter left in me. At 50, I was relatively young. In good shape. In good health. Plus, I loved the feeling I got fighting a fire, the adrenalin rush mixed with power and purpose. The thrill of the effort. The cheating of death. The frat-house camaraderie. The whole damn thing. And I loved the way the job left me looking in the eyes of my children. Some weeks later, Lisa told me how the children of firemen place their fathers on pedestals, how they grow up knowing that when people talk about how firefighters are ordinary men made extraordinary by acts of bravery it's a kind of sham, because the truth is firemen are born extraordinary people. Imagine hearing something like that from your own kid. So, yeah, I loved the whole package, the good and the bad, and it hadn't occurred to me that it was time to punch out. But there I was, on the phone with my wife, talking about quitting. Like I said, it hadn't even occurred to me, in any kind of front-burner way, but I'd just been through hell, and it kinda spilled out of me, all on its own.

'Pitch,' Debbie said, incredulous, 'what are you talking about?'

'I don't know,' I said. 'I guess I'm just talking. I'm thinking maybe I've had enough.'

Debbie cried – again, or still, I couldn't tell. And we moved on to other things. Small talk mostly. How the kids were doing. Who had called. What was going on in the department. The other wives she'd spoken to. She wanted me home, and soon, and a part of me wanted the same, but I told her I had to make my way to the firehouse first. We'd lost so many guys, too many to count. There were widows to call, brothers to comfort. The sad business of mourning was all around, and I needed to be in its middle. I was the chief. It was part of the job, but more than that it was a part of who I was.

I may have been done, but I wasn't done just yet.

AFTERMATH: The Site

For the longest time, I couldn't bring myself to go back down to the World Trade Center site. Almost immediately the place had become the focus of this enormous outpouring of grief, patriotic spirit and volunteerism. It was like a magnet for every conceivable human response to this immense national tragedy. Firefighters drove clear across the country, on their own time and their own dime, to pitch in to the rescue and recovery effort. Reporters and photographers flew in from all over the world to record the scene. Every conceivable charitable organization (and, frankly, some inconceivable ones as well) lay some kind of claim to the site. World leaders and well-connected celebrities took turns posing with New York Mayor Rudy Giuliani to publicly demonstrate their connection to the tragedy. Tourists and rubberneckers pressed as close to the complex as the yellow police barrier tape and sawhorse barricades would allow. Without even realizing it, folks were drawn to that sixteen-acre complex, and I suspect they will continue to be pulled there for as long as we remember the events of that day.

For New Yorkers, strangely, a visit to the site became a kind of rite of passage, through which we took turns accepting what had happened, and what was happening still. In some circles, the perception was that you had to go down there to understand the devastation. To truly *know*. For a lot of guys in the department, logging those long hours looking for our lost brothers was a kind of penance, a way to ease the survivor's guilt, a place to put a whole bunch of bottled-up and uncertain emotions. The site came to stand for so many different things, for so many different people.

As for me, though, I just couldn't go. A part of me desperately wanted to be down there, but that part was out-voted by the rest of me. In the first weeks following the attack, my eyes were in such seriously bad shape that I couldn't be outside in direct sunlight for any length of time, or anywhere near the kind of acrid smoke that surrounded the area, but I realize now that in a lot of ways these were just convenient medical excuses to keep me away. There was a more fundamental reason, too, one that would still prevent me from visiting the site once my eyes improved and the smoke cleared: it was too much, too soon, too close, too raw. It was too … everything. And I was too frazzled to have my nose rubbed in what I'd just been through so soon afterwards. I wasn't ready for it.

By the time I got back to the firehouse on the night of September 11th, the folks on the news had already dubbed the World Trade Center site 'ground zero', and the term made me bristle the moment I heard it. It rang false, and trite, and somewhat off the mark. *Ground zero?* Listening in to the Dan Rathers and Tom Brokaws of the world, the term just rubbed me the wrong way. It sounded as if the World Trade Center complex itself was the source of the terrorist attack on our freedom, on the men in my command, on the thousands of people who worked in those towers; as if all of us here at home were somehow responsible for such a monstrous, unthinkable act of violence. Maybe it was just me reacting in this way. I don't know. I never talked about it with anyone because it seemed like a small thing. Big enough that I refused to use the phrase *ground zero* in the original manuscript of this book, and big enough that I avoided it in my private conversations, but not big enough that I needed to make a huge deal out of it if it came up on its own. In fact, I once heard the phrase from the mouth of a priest at a memorial service who confessed that he didn't like it either; he felt the place was sacred ground and shouldn't be reduced by the negative connotation.

It's interesting, though, that firefighters and other relief workers never embraced the phrase. To them, the World Trade Center site became known as 'the pile', and that seemed to me more appropriate. That's what it was: the left-behind pile of a great symbol of American ingenuity and commerce, and a fresh burial ground for way too many innocent victims. Something to work to get to the bottom of. A place to stand, undaunted. (After the pile disappeared, the area became known among rescue workers simply as 'the site', which I also thought was fitting.)

By any name, I still couldn't bring myself to make the trek downtown. I don't know what I was afraid I'd find, or even if I was afraid at all, but I was totally unmotivated about it. I had this big-time inertia. I'd get it into my head to go, and then I'd find a reason not to. God knows, I didn't want for excuses: doctor's visits, fire-department paperwork, visiting the widows of the men in my battalion, dealing with the strain of emotions in my own home. Around the time of the first few fire-department memorial services, however, I started to think more and more about traveling down there, and began to confront some of these fresh memories head on. Beginning in late September, and stretching pretty much throughout the rest of the year, there was a constant, endless string of funerals and memorial services – and, as a chief, I wanted to attend as many of these as I could. Certainly, I had genuine, lifelong friendships with a ton of guys who were lost on September 11th. I lost men in my command, guys I'd worked with over the years, guys who'd worked under me, guys I'd studied with, trained with ... you name it. At one point, I was going to six or seven of these services a week. The department did what it could to coordinate this effort, but when you have to pay last respects to three hundred and forty-three individuals, there aren't enough days on the calendar to do so one at a time.

At each service, there was a new layer of grief and memory to get past, a deepening sense of clarity about what had happened, an encroaching dread over what might happen next. Fatherless children, wives missing their husbands, parents missing their children ... there was a great weight in each and every chapel, and on top of that there was the emotional uncertainty of the world around. Think back to the early part of the fall and you'll remember it as the time when the New York area was gripped by fears of follow-up attacks, when post offices and media outlets around the country were dealing with anthrax scares, and when airport security measures made routine travel a nightmare. It was an impossible time, and to add to this mix I was having my own troubles dealing with the aftermath of the collapse of those towers. I couldn't sleep at night. I couldn't look at my wife or my children without placing myself back in that black, smoke-filled void. I couldn't stop hearing the cascading roar of those towers collapsing. I couldn't chase the real and imagined terrors from my racing imagination. I couldn't carry on a normal conversation without drifting back in my mind to the events of September 11th. Hell, I couldn't even sit through a two-hour movie without finding something on the screen that would take me back to those harrowing moments just before the collapse.

The department set me up early on with someone to talk to, a therapist who understood about things like post-traumatic stress syndrome. Me, I didn't have the first clue about anything a professional person might call a *syndrome*, but it didn't take a therapist to get me to realize I needed to shake things up and confront them head-on if I meant to get past them. All it took was time, and maybe a little bit of perspective, and by mid-October I was ready to go down to the site and see what I might find there. I convinced myself it was a working visit – I wanted to bring my collaborator on this book to the shell of the B stairwell, so he could get a good visual on it and help me to write about it – but I knew it was more than that. I knew it was me giving myself the push I needed to get past a difficult hurdle.

What I found down there was a revelation – and nothing at all like I expected. First of all, the topography of the place had undergone a sea change in just a few short weeks. What had only recently been acres upon acres of twisted steel, shattered glass and mountainous rubble was now a fairly smoothed-over construction site. At least, that's how it looked in parts. There was a still a considerable 'pile', big enough that I couldn't fathom how all the construction crews in the world could sift through it for body parts and personal effects and still manage to cart it away, but next to it there was an impossibly vast stretch of space that had already been cleared. Less than a month after the collapse of the two tallest towers in the world, the plaza below resembled a construction site more than it did a disaster site. The area of a good football field or so had been bulldozed over with earth, to give the big trucks and heavy machinery the necessary room to maneuver. Underneath, of course, there were still unimaginable tons of rubble to contend with, but to the naked eye the biggest piece of the puzzle had already been filled in.

The shell of my old stairwell was long gone, and I struggled to place it. Like I said, there was still a 'pile', but it was nowhere near the scope and scale of the vast debris field that awaited us on that black afternoon. I could get a rough idea where the staircase opened up onto that shorn-off balcony, fixed against some of the still-standing perimeter buildings, but I couldn't pinpoint the area with any kind of precision.

Surprisingly, I wasn't overcome with any watershed of emotion. I saw a bunch of guys I knew, and it was a welcome thing to be collected into their arms with cries of 'Hey, Pitch, it's great to see you!' and 'I heard about what happened, Chief, glad somebody made it out okay.' In the extra efforts of these good men was the spirit of the fire department, the spirit of the city, the spirit of the country. We would get past this thing, whatever it took. We would find a way to move this pile, and bury our dead, and rebuild our lives. It was a beautiful afternoon, as I recall, an Indian summer sort of day. In shirtsleeves, with a high sun and virtually no wind, I walked about that plaza, trying to get my bearings against the still-hanging footbridge, thinking it was a lot like the beautiful morning of September 11th, before everything turned to shit. Clear skies. Warm temperatures. Endless possibility. And now here we were, on the other side of a kind of hell, and it was still pretty damn nice out, and there were still smiling faces happy to see me, and I was able to smile myself, more than I had thought possible, and there were reminders everywhere I looked that life does indeed go on, that the human spirit is a tough thing to kill.

I went back again just a week or so later, and again a week or so after that, and each time it was a little bit easier. Tougher, in a bigger-picture sort of way, but easier on a personal level – and by that I mean I was growing more and more used to the ordeal we had suffered in that stairwell, and more and more confounded by the enormity of the tragedy everywhere else on that plaza. The clear, terrifying memory of what happened in those towers never faded, but I managed to slide it to that place in my mind where it wasn't exactly front and center. Gradually, I was able to look at the site like my colleagues on the line, assess the work that needed to be done, anticipate the areas prone to secondary collapse, calculate the man-hours needed to do this or that piece of work. I moved from thinking of myself as a victim or a survivor, to thinking of myself as someone whose job it was to assess the damage.

This shift in my own outlook was never more apparent than on the afternoon of November 6th, when I went down to the pile with my wife Debbie. It was to be her first visit to the site, and we were both a little anxious about it. You know, it's funny, but we did a lot of talking around the issue in the weeks just after September 11th. A lot of what we both wanted to say to each other kinda hung in the air between us. It was almost like each of us knew what the other was thinking, or like we wanted to have that connection of understanding, but it was too painful to give these thoughts real voice. And so we talked around things. We left things unsaid. We took turns assuming we knew what was on each other's minds.

We didn't talk much, as we approached the debris field. There were tears in Debbie's eyes, as I recall, but mostly we were quiet. There wasn't much to say. Every now and again we'd stop to talk to someone from the department, someone who hadn't seen us since the day of the attacks, but when we were alone, just the two of us, we silently took in the scene.

While we were standing by one of the department command posts, there was a great commotion over at the centre of the field. From where we stood, it looked like there had been a collapse of some kind. Debbie actually saw a fireman disappear from her line of sight, and shouted, 'Pitch, that fireman just disappeared!' Almost immediately, rescue workers started running in every direction, trying to place the source of the confusion, trying to figure out what had happened. Of course, I was dressed in my civilian clothes – leather jacket, jeans, blue golf shirt – and nobody knew who I was, so there was nothing for me to do but take in the unfolding scene, and re-imagine myself in the middle of an eerily similar situation. At one point, after the commotion had died down somewhat, and the other firefighters had a good bead on rescuing this guy, I inched over to the opening. I was astonished to see this enormous drop all of a sudden, out of nowhere. It was like there was a sudden ledge, dropping thirty feet and opening into an area about the size of a small gymnasium. The guy down below was banged up a little bit, but mostly fine, buried in this void like we had been just a couple of months earlier. Not for anywhere near as long, and not under anything like the perilous, terrifying conditions we had to face on September 11th, but close enough that it got my heart racing a little bit. And forget *my* heart – this guy's heart must have been pumping pretty good too at this point.

They managed to drop a ladder down to him soon enough, and he managed to climb out, and the work continued, and from what I hear it was the only incident of its kind during the entire recovery effort. There were some minor secondary collapses, but there was no bottom-dropping accident of this magnitude, and as far as I know it never made the papers. There's an incident report on it, you can be sure about that, but most people don't know about it, and the reason I mention it here is for the way it kinda threw me back those couple of months into the shell of that interior stairwell.

And Debbie! Man, I'd forgotten about Debbie in the few minutes it took for this incident to play out! She was fairly freaked out about it, which I could understand. Just as my thoughts ran to me and the guys and Josephine in that stairwell, so did Debbie's too, and her thoughts shook her up pretty good. When I returned to the spot where I'd told her to wait for me, she had this white, ashen look on her face. On the surface, the look might have been for this one individual who had narrowly escaped danger in this one near-miss, but underneath I knew it ran to some other place. To me. To what we had both nearly lost on September 11th.

Here again, we didn't talk about it. Not directly. We came at it in our sidelong way. I put my arms around her, and she put her arms around me, and we held each other. Right there, in the middle of this strange sea of activity, we just held each other tight, and willed the rest of the world away.

'It's gonna be okay,' I finally said. 'It's gonna be okay.'

ACKNOWLEDGEMENTS:
The Power To Lead

I've been told it's the custom, at the beginning or end of these books, for the writer to offer up a couple of lines of thanks and praise to the many industry types who had a hand in the project. Ghostwriters, researchers, agents, publishers, editors, publicists... all down the line. Even friends and family members get a kind word (deserving or not). As a reader, this always strikes me as a little bit too 'inside'. As a writer, now, it seems beside the point. I mean, we're all professionals, right? But I'm a firefighter, not a writer, and there's little chance I'll run into the good people backing this project once it runs its course; we've all had our jobs to do, we've all meant to do them well, and I sign off on these pages knowing we've all succeeded. I'm grateful to each and every one of you, for each and every one of your contributions and suggestions. I'm grateful to my wife and children, mother, father, my Aunt Jeanne who is like a second mother to me, my brother Robert, and all of my cousins, also the FOF.

But I have a different sort of acknowledgement in mind, and I'd like to bend the custom to suit my purpose, to fill these next pages with thanks and praise for my friends and mentors in the fire department, because if you've read this far with me in the book, you'll know that I wouldn't be here without their strong example.

First, a final word or two on the Fire Department of New York and our unique chain of command. As I've written, we're like a military organization, in structure and practice, and we really do look at every job as a kind of going into battle. These aren't empty words, they're a fact of our working life. And, as in the military, we live and die by our leaders. Throughout the department, throughout my career, there have been strong-minded, strong-willed, strong-hearted deputies, chiefs, lieutenants, captains and senior firefighters to guide me, support me, and cover my ass, and it's to these strong individuals that I am truly grateful, to the senior firefighters most of all. In the good old days, when I was just starting out, it seemed to me there was incredibly strong leadership, at every level of the department, and as I've moved up in the department I've noticed that the ranks have thinned. I guess this has to do with a shift in my perspective. After all, when you're a rookie, everyone who's been around looks like a pro, and when you're a pro, everyone else looks like a rookie. So there's some of that distorting my view, but I think it goes to something deeper, and as long as I have the floor I think I'll vent for a bit. Each firehouse has 25 firemen, three lieutenants and one captain, and at any given time you'll find four or five firefighters and one officer on duty. When I joined the department, there were anywhere from five to seven firefighters on duty each tour, but over the years the bean-counters in administration have cut that number to present levels – and in the process, cut at the heart of what we do. Battalions that once supervised four or five companies are now in charge of six to ten. At one point, there were 25 staff chiefs overseeing the daily operations of the entire department; as of writing this, that number has dwindled to nine. And the cost-cutting doesn't end here. We move about in the Dark Ages, fighting for low-cost items and basic services. We're made to live and eat and sleep in the firehouse for long stretches, and expected to stock and outfit our own kitchens and bathrooms. This is a small thing, I know, but to some of these guys on the line, it's everything. Show me a salaried professional who has to bring his own toilet paper if he wants to take a shit at the office, and I'll show you a labour movement waiting to happen. Did you know that in over three-quarters of all firehouses, it's impossible to make an outside telephone call unless it's from the chief's office? How ridiculous is that? We ask our bosses to play secretary and phone tag, and make it all but impossible for a captain or lieutenant to call back one of his men without using a payphone.

Let's face it, the tighter we are on money-matters, the harder it is for firefighters to do their jobs, and for the true leaders – not the brass upstairs, but the guys on the line – truly to lead. These are the informal leaders of the department, and without them we couldn't function. Sadly, there are too few of them still around, as our ranks dwindle, but they're out there, in every firehouse. The no-bullshit, non-officers who break in the probies, teach them 'the right thing to do', push them down the right path and knock some sense into them when they screw up. Lieutenants, captains and chiefs are not in a position to do such hands-on mentoring, but the senior fire-fighters do the trick, and they do it well. And so, for the jump-start they gave me in my own career, I offer a tip of the helmet to the senior men of Engine Co 91.

Jim McCloskey, Al Quinn, Jack McCarthy and Bob Schildhorn, in the beginning. And later on, as I got promoted and moved on to other houses, Tom McCarney, Jim McCoy, Tom Coates, Richie Polizzio, John Greehy, Vinnie Albanese, Bob Hagan, Joe Brosi, Pete Farreenkopf, Jerry Nevins, John O'Donnell, Al Miskiewicz, Walter Dryer, Willie Martinez, George Joos, Dan Murphy, Larry Fursich, Joe Pigot, John Haggerty, Joe Fallon, Cris Donavan, Jim Shatz and Tim Lipenski. I know I'm leaving out hundreds of good soldiers and solid leaders, but I wanted to be sure to mention these few brothers who had a firm hand in shaping the firefighter I would become.

I've worked with more than a few good officers, and they also rate a mention here: Walter Brett, Nick Visconti, Tom Galvin, Dennis Collopy, John Hughes, Pat McNally, Peter Incledon, Jack Rynne, Paul Jirack, Dan Loeb, Gary Senger, John Pelligrinelli, Rich Toban, the Meehan brothers, Dan Twomey, Joe Guglielmo, Tony Palazzola, Phil Curran, John Byrnes, Jerry Reilly, Steve Baker, Charlie Bonar... I could go on and on. Dave Wooley, Audie Meagh, J.J. Johnson, Tom Roberts, Kevin Guy (the funniest man in the department), Ed Kilduff, Jim Hodgens. Really, I could name just about every member of the 11th Battalion and the Third Division, but these guys know how I feel about them without seeing their names in print. And how I feel is this: we're all born leaders. It's what drives us to the job in the first place. We're leaders in our families, and leaders in our communities. We give a shit. (And we buy our own toilet paper!) And we refuse to let the cowardly bottom lines of our administrators get in the way of *our* bottom line – saving lives, watching each other's backs, doing what needs to be done. And so, to those mentioned here, and to those whose names I carry in my heart, I am grateful. You might not know it, but you all had a hand in hauling my ass out of that stairwell and down to safety. Without you, I'm nothing.

'I'll see you at the big one.'

Index